SpringerBriefs in Applied Sciences and Technology

SpringerBriefs present concise summaries of cutting-edge research and practical applications across a wide spectrum of fields. Featuring compact volumes of 50 to 125 pages, the series covers a range of content from professional to academic.

Typical publications can be:

- A timely report of state-of-the art methods
- An introduction to or a manual for the application of mathematical or computer techniques
- A bridge between new research results, as published in journal articles
- A snapshot of a hot or emerging topic
- An in-depth case study
- A presentation of core concepts that students must understand in order to make independent contributions

SpringerBriefs are characterized by fast, global electronic dissemination, standard publishing contracts, standardized manuscript preparation and formatting guidelines, and expedited production schedules.

On the one hand, **SpringerBriefs in Applied Sciences and Technology** are devoted to the publication of fundamentals and applications within the different classical engineering disciplines as well as in interdisciplinary fields that recently emerged between these areas. On the other hand, as the boundary separating fundamental research and applied technology is more and more dissolving, this series is particularly open to trans-disciplinary topics between fundamental science and engineering.

Indexed by EI-Compendex, SCOPUS and Springerlink.

Marcus Vinicius da Silva Neves ·
Gustavo Arruda Ramalho Lira ·
Guilherme Fortes O. F. Lima

Offshore Operations Engineering

A USeful Guide for the Offshore Worker

Springer

Marcus Vinicius da Silva Neves
Petroleo Brasileiro SA (PETROBRAS)
Rio de Janeiro, Brazil

Gustavo Arruda Ramalho Lira
Petroleo Brasileiro SA (PETROBRAS)
Rio de Janeiro, Brazil

Guilherme Fortes O. F. Lima
Petroleo Brasileiro SA (PETROBRAS)
Rio de Janeiro, Brazil

ISSN 2191-530X ISSN 2191-5318 (electronic)
SpringerBriefs in Applied Sciences and Technology
ISBN 978-3-032-04714-4 ISBN 978-3-032-04715-1 (eBook)
https://doi.org/10.1007/978-3-032-04715-1

This Springer imprint is published by the registered company Springer Nature Switzerland AG
The registered company address is: Gewerbestrasse 11, 6330 Cham, Switzerland

Foreword

This book makes a notable contribution to the technical literature on offshore asset management. It is a unique and valuable resource because it addresses a subject that, until now, has not been covered in a single, consolidated volume. From my experience in the industry, I have observed gaps in the professional training of many practitioners, which this book is uniquely positioned to help fill. It provides a clear and consistent treatment of complex themes, including life cycle management, integrity, operational safety, life extension strategies for offshore installations, and emergency response. To my knowledge, no comparable reference exists that brings these topics together in such a coherent way.

The content is structured to be a useful guide for professionals in the field while also serving as a solid academic reference. Its design, which includes links to relevant articles, standards, and publications, allows the reader to go beyond the text and explore the underlying technical and scientific foundations in greater depth. This balance between conceptual discussion and applied methodology makes the work a valuable resource in academic, corporate, and operational environments.

A defining strength of this book is the complementary expertise of its authors. Their distinct professional experiences and academic backgrounds converge to form a work that is both rigorous and accessible. This combination supports strategic decision-making in the industry and can inspire new research. It is at this intersection of practice and theory that the book provides its most relevant contribution.

In my assessment, this is a necessary and timely work. It will serve as a key reference for those who seek to understand and apply best practices in the management of offshore installations throughout their life cycle.

Guilherme Sargenti
General Manager
Petrobras' Espírito Santo Operational Unit
Vitoria, Brazil

Acknowledgements

In the amazing process of writing *Offshore Operations Engineering: A Useful Guide for the Offshore Worker*, cooperation and support played a crucial role. This book, a synthesis of engineering knowledge and real-world experience, is a tribute to the hard work of many people, for whom we are eternally grateful:

To our students, colleagues, and friends in the field, thank you for your insights and shared experiences, which helped strengthen the quality and relevance of this work.

We are also grateful to our families—for the guidance and example of our parents, the support and patience of our spouses, and the inspiration brought by our children.

Finally, we acknowledge the offshore operations engineering community, whose dedication and ongoing contributions to the discipline continue to inform and inspire our efforts.

With deep gratitude and modesty,

Marcus Vinicius da Silva Neves
Gustavo Arruda Ramalho Lira
Guilherme Fortes O. F. Lima

Contents

About the Authors

Marcus Vinicius da Silva Neves has been an Equipment Engineer at PETROBRAS since 2002. He holds a D.Sc. in Energy Planning and an M.Sc. in Mechanical Engineering, both from the Federal University of Rio de Janeiro (COPPE/UFRJ). Certified as an Energy Manager by the American Society of Energy Engineers (AEE), he has held various managing positions at PETROBRAS, including Maintenance Engineering Manager, Operational Safety Manager, and Offshore Installation Manager. Currently, he serves as the Manager of the Center of Science and Technologies related to Equipment and Process Engineering at Petrobras University. He is the author of the books *Energy Efficiency in Oil Production* and *Thermoelectric Power Plants*. Additionally, he has served as a Collaborator Professor at Fluminense Federal University (UFF) and Rio de Janeiro Federal University (UFRJ), among others, for post-graduate programs.

Gustavo Arruda Ramalho Lira is an accomplished petroleum engineer with over 20 years of experience in the E&P industry. His expertise spans well engineering, drilling, completion, and workover operations on onshore and shallow-water wells, as well as production operations in ultra-deep-water FPSOs focusing on Brazil's pre-salt reservoirs. Additionally, he has developed strong managerial skills, holding leadership positions for the past 15 years. He holds a degree in Civil Engineering and a Master's degree (M.Sc.) in Urban Engineering from the Federal University of Paraíba (UFPb). In the business field, he completed an MBA in Business Management at Fundação Getúlio Vargas (FGV) in 2016, the Harvard ManageMentor Program in 2021, and a postgraduate specialization in Management at Fundação Dom Cabral (FDC) in 2025. Since joining Petrobras in 2004, Gustavo has held key managerial roles, including Critical Resources Coordinator, Well Engineering Manager, and, for the past eight years, Offshore Installation Manager aboard the P48 and P74 FPSOs. In academia, he has been an adjunct professor at the Federal University of Rio Grande do Norte (UFRN) since 2009, teaching in the Petroleum Engineering Department.

Guilherme Fortes O. F. Lima has been with PETROBRAS since 2007. He graduated in Naval Engineering in 2004 and holds an M.Sc. in Ocean Engineering, both from the Federal University of Rio de Janeiro (UFRJ/COPPE). Throughout his career at PETROBRAS, he has held various important positions primarily related to offshore production on oil platforms. He worked onboard as a Barge Supervisor and Offshore Installation Manager on two major installations and later took on the role of Operational Manager for another major production platform. Currently, he manages a research and innovation team focused on safety at PETROBRAS' research center (CENPES).

Chapter 1
Introduction

Abstract Offshore Operations Engineering: A Useful Guide for the Offshore Worker offers an accessible, practice-oriented overview of offshore oil production for engineering students, offshore crews, and technical/management support teams. Framed around safe, efficient, and legally compliant operations, the book explains the harsh metocean environment and its implications for structural design, stability of floating units, and day-to-day operations; reviews essential navigation skills for support, rescue, and lifeboat handling; clarifies the regulatory landscape spanning maritime, environmental, and industry authorities; and adopts a full life-cycle lens—from planning and construction through pre-operation, operations, maintenance, and end of service. Core themes include process and occupational safety, emergency management, and operational risk control, reinforced by ready-to-use guidance. Written in clear language, it equips readers to understand the scope and interdependencies of offshore work and to apply sound engineering and management practices on board.

The purpose of the book "Offshore Operations Engineering: A Practical Guide for the Offshore Worker" is to provide engineering students and all those who work directly in offshore oil production facilities or provide technical and management support for these units with an overview of all issues that are related to the project; as well as safe, efficient, and legally compliant operation. This book was written with the intention of being a practical guide for offshore workers.

It is not an easy undertaking to operate a production facility that is located offshore. In addition to all of the concerns pertaining to the manufacturing process, maintenance, integrity, operational safety risk, and profitability of a common industrial unit, there are a few other things to take into consideration. To begin, there is the unfriendly and hostile environment, which is characterized by currents, waves, and winds that are both intense and consistent, representing a much more challenging environment than that found on dry land. In addition to this, the environment is highly corrosive. Oil production platforms need to have their structures well dimensioned to support these loads simply because they operate at sea. Additionally, in the case of floating installations, these structures need to remain stable in order to allow the operation to

M. Vinicius da Silva Neves et al., *Offshore Operations Engineering*,
SpringerBriefs in Applied Sciences and Technology,
https://doi.org/10.1007/978-3-032-04715-1_1

make safe and efficient use of its production, utility, and habitability systems. This is especially important because of the environmental efforts that were mentioned earlier. The expert who works in the production of oil at sea needs to have at least a basic understanding of the structure of these facilities, in addition to having an understanding of the fundamentals of stability.

Working at sea needs the crew to not only know how to control the support vessels that serve their unit but also how to pilot rescue boats and lifeboats. This is in addition to knowing how to lead the vessels that provide supplies to the unit. As a result, having a solid foundation in navigational skills is essential.

Legal compliance is another very important element to consider. There are a great number of national and international regulations, some of which are associated with the following: the national maritime authority, the maritime platform flag authority, the international maritime regulation, the environmental agency, the regulatory agency for oil production activity, the classification society, helicopter operation, telecommunications, chemical products, surveillance health, and many others. As a result of this, it is essential for the worker offshore to have a comprehensive understanding of the regulations that involve maintaining legal compliance during the extraction of oil at sea.

In addition, it is necessary for us to have an understanding of the various components that are associated with the life cycle of these facilities, beginning with the planning and construction phase of the unit, continuing through the preparation for operation phase, the operation phase, and the maintenance phase, and finally ending with the conclusion of its productive mission. In light of this, it is essential that we are familiar with the ideas of safety and emergency management in addition to knowing how to manage the operational safety risk associated with this kind of activity that is exposed to a variety of risks.

We have prepared practical examples of some topics that have been taught throughout the book as a way of consolidating the knowledge that has been received and contributing to the fixing of it. As a result, the readers will have access to a useful reference that they can refer to throughout the course of his everyday activities.

We hope to accomplish this by providing a reading that is not very difficult and is straightforward, making it available to anyone who has a desire in enhancing their understanding of offshore operations engineering. We have great hopes that by the time the reader is finished with this work, he will have his vision honed enough to be able to grasp the scope of the issue as well as the myriad of facets that are involved in the offshore oil production.

Chapter 2
Nomenclature of Naval and Oceanic Engineering

Abstract The chapter "Nomenclature of Naval and Oceanic Engineering" introduces the fundamental terminology used in naval and ocean engineering, essential for the design, construction, and operation of marine vessels and offshore structures. It describes the main structural components of a ship—such as the hull, bow, stern, deck, bulkheads, rudder, propeller, and keel—along with their reinforcing elements and functional roles in ensuring structural integrity and seaworthiness. The chapter also explains the six degrees of freedom that characterize a ship's motion at sea (heave, sway, surge, roll, pitch, and yaw), distinguishing between linear and rotational movements. This technical vocabulary forms the foundation for clear and precise communication among naval engineers, designers, and offshore operators.

Marine structures and vessels are designed, built, and maintained by naval and marine engineers. To explain the numerous components of a ship and its motions, this field uses a distinct set of terms. For experts in the field, understanding this terminology is essential. With a focus on the nomenclature of ship construction and the primary movements of vessels, this chapter provides an introduction to the fundamental terms used in naval and maritime engineering.

Nomenclature of Ship Structures

1. **Hull**: The main body of the ship that floats on the water. It includes the bottom, sides, and deck of the ship.
2. **Bow**: The front part of the ship.
3. **Stern**: The rear part of the ship.
4. **Portside**: The left side of the ship when facing forward.
5. **Starboard**: The right side of the ship when facing forward.
6. **Engine Room**: A compartment in which the main propelling equipments and machines are arranged.
7. **Deck**: The top surface of the ship's hull that forms the roof of the hull's structure.
8. **Bulkhead**: A transverse section plate that works as a partition separating ship compartments, and giving it also structural resistance.

M. Vinicius da Silva Neves et al., *Offshore Operations Engineering*,
SpringerBriefs in Applied Sciences and Technology,
https://doi.org/10.1007/978-3-032-04715-1_2

9. **Rudder**: A flat piece of material attached with hinges to the ship's stern, used for steering.
10. **Propeller**: It is a rotating fan-shaped structure that propels the ship using power from the main engine. It converts rotational motion into thrust, driving the ship through the water.
11. **Scupper**: A hole in the side of a ship at deck level that allows water to be drained back overboard.
12. **Mast**: A tall vertical structure on a ship, serving as a support for sails, spars, and rigging.
13. **Bridge**: The location from which a ship is commanded, typically located on the upper deck from the ship's superstructure.
14. **Shell Plating**: This refers to the metal sheets covering the hull, deck, and bulkheads of the ship. It is the outer skin of the ship, and provides watertightness and resistance to the water external pressures.
15. **Sheer Strake**: The uppermost edge strake of the side shell plating, which is typically thicker and stronger than the other strakes around it.
16. **Stringer Strake**: The outer edge strake of the deck shell plating, which is typically thicker just as the sheer strake is.
17. **Bilge Plate**: The bilge plate is the curved plating located between the bottom shell and the side shell, where the bilge keel is typically welded as a fitting to reduce the ship's tendency to roll.
18. **Longitudinal Hull Girder**: A longitudinal primary structural members installed on the center bottom of the hull. They provide primary strength and longitudinal stiffness to the ship's structure.
19. **Keel**: A centerline bottom structural member around which the hull of a ship is built. It runs along the centerline of the ship, from the bow to the stern, where the frames are attached on each side. The keel is often a girder of high dimensions.
20. **Side Girders**: These are longitudinal primary structural members installed on the side of the hull.
21. **Transverse Side Frames**: These are transverse primary structures, like ribs, that provide the ship with its shape and strength. They extend from the keel to the upper deck and are typically spaced evenly along the ship's length.
22. **Bottom Transverse**: A transverse primary structural members installed on the bottom of the hull. They provide primary strength and transverse stiffness to the bottom structure.
23. **Deck Girder**: A continuous stiffening member which runs fore and aft along a ship to support the deck.
24. **Deck Transverse**: A transverse primary structural members installed on the deck of the hull. They provide primary strength and transverse stiffness to the deck structure.
25. **Bulkhead**: A primary vertical partition structure that separates compartments within the hull, such as cargo and ballast tanks, and contributes to the overall structural stability. It can be either transverse or longitudinal, as for example the centerline bulkhead.

26. **Hatch**: A hatch is an opening in the deck of a ship or the floor of a warehouse. Hatches provide access to the interior and are typically covered by a watertight door or hatch cover.

Main Movements of Vessels

A ship at sea moves in six degrees of motion: heave, sway, surge, roll, pitch and yaw. The first three are linear motions, and the others are rotational around the three axes.

Rolling is a rotation around a longitudinal axis, pitching is a rotation around the transverse axis and yawing is a rotation around the vertical axis.

1. **Heave**: The vertical movement (up/down) of the ship along the vertical Z-axis.
2. **Sway**: The lateral movement (side to side) of the ship along the transverse Y-axis.
3. **Surge**: The longitudinal movement (forward/backward) of the ship along the longitudinal X-axis.
4. **Pitch**: The rotation of the ship around a transverse Y-axis.
5. **Roll**: The rotation of the ship around a longitudinal X-axis.
6. **Yaw**: The rotation of the ship around a vertical Z-axis.

Bibliography

1. Naval History and Heritage Command, *Nomenclature of Naval Vessels* (2025). https://www.history.navy.mil/research/library/online-reading-room/title-list-alphabetically/n/nomenclature-naval-vessels.html. Accessed 18 Apr 2025. This source from the U.S. Navy History and Heritage Command provides a nomenclature of naval vessels
2. Royal Navy, *The Naval Engineer* (Royal Navy, London, 2012). https://www.yumpu.com/en/document/read/48823146/the-naval-engineer. Accessed 15 Apr 2025. This document from the Royal Navy discusses various topics relevant to naval engineering
3. The Society of Naval Architects and Marine Engineers, *Principles of Naval Architecture Second Revision* (SNAME, Jersey City, 1989). https://ftp.idu.ac.id/wp-content/uploads/ebook/tdg/ADNVANCED%20MILITARY%20PLATFORM%20DESIGN/Principles%20of%20Naval%20Architecture%20(Second%20Revision),%20Volume%20III.pdf. Accessed 14 Apr 2025. This book provides a comprehensive overview of the principles in the field of naval architecture
4. United States Naval Academy, *Naval Architecture & Ocean Engineering* (2025). https://libguides.usna.edu/c.php?g=410506&p=2795318. Accessed 14 Apr 2025. This guide from the U.S. Naval Academy provides resources related to naval architecture and ocean engineering
5. D.G.M. Watson, *Practical Ship Design* (Elsevier, Amsterdam, 2002). This book is part of the Elsevier Ocean Engineering Book Series and discusses practical aspects of ship design

Chapter 3
Offshore Oil Production Installations

Abstract The chapter "Offshore Oil Production Installations" presents an overview of the main types of offshore oil and gas production systems, describing their structural configurations, functional components, and operational principles. It classifies offshore facilities into fixed, floating, and subsea systems, explaining how factors such as water depth, environmental conditions, and field characteristics determine the most suitable installation type. The chapter also introduces key elements of production systems, including subsea equipment, risers, flowlines, and topside modules, emphasizing their integration for efficient hydrocarbon processing. Additionally, it highlights the importance of safety management during Simultaneous Operations (SIMOPS), where multiple offshore activities occur concurrently, requiring strict coordination and risk control. Overall, the text provides a concise yet comprehensive foundation for understanding the engineering, operation, and safety aspects of offshore production facilities.

This section provides an overview of key aspects of offshore oil production facilities, aiming to acquaint the reader with their unique characteristics.

Although several structures and facilities operate collectively for oil production, they can be categorized, for educational purposes, based on their position relative to the seabed, as shown in Fig. 3.1.

Two important structures for positioning floating platforms are not included in Fig. 3.1 because they are not directly involved in the oil and gas production process. These structures are the anchors and mooring lines or cables, located in the subsurface and on the seabed, respectively. References to these structures can be found in Ma et al. [20].

The subsurface structures consist of production and injection wells. Constructing an oil well capable of fulfilling its production or injection purposes typically involves two stages: drilling and completion. Drilling creates access from the surface to the reservoir, while completion prepares this access for safe and economical production or injection.

In shallow water, these wells are usually constructed using jack-up rigs. In deep water, above 2000 ft., floating rigs such as semi-submersibles or drillships are

M. Vinicius da Silva Neves et al., *Offshore Operations Engineering*,
SpringerBriefs in Applied Sciences and Technology,
https://doi.org/10.1007/978-3-032-04715-1_3

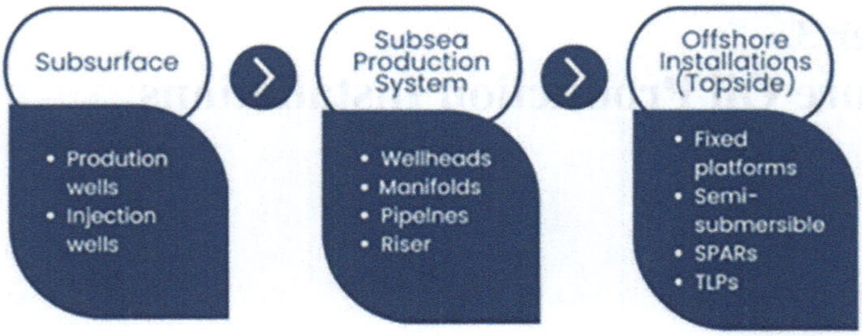

Fig. 3.1 Offshore installations and structures

employed, with the latter being predominant in ultra-deep water (over 6000 ft.). Modern drilling rigs are equipped with dynamic positioning systems, which maintain position and heading through thrusters that counteract environmental forces such as waves, currents, and wind, both in the intensity and direction of the applied force [15]. IMO has published a guide with recommendations on design, equipment, operation, and testing, aimed at reducing risks to people, vessels, and structures when operating under dynamic positioning [16].

The construction process of offshore wells, the types of rigs used and their subsystems are beyond the scope of this publication and can be found in other research sources [1, 19, 22]. In this chapter, the structures of subsea systems and offshore production installations will be detailed. Additionally, important aspects of simultaneous operations (SIMOPS) will be discussed, which typically involve these three groups of structures and installations: wells, subsea systems, and topside.

3.1 Subsea Systems

API Recommended Practice 17A provides "general requirements and recommendations for the development and operation of subsea production/injection systems, from the concept development phase to decommissioning and abandonment" [4, p. 1], and establishes that "a complete subsea production/injection system comprises several subsystems necessary to produce hydrocarbons from one or more subsea wells and transfer them to a processing/host facility located offshore (fixed, floating, or subsea) or onshore, or to inject water/gas via subsea facilities and/or wells" [4, p. 1]. Within this series of recommended practices (API RP 17), API outlines crucial requirements for these systems, including reliability and risk management, equipment qualification, safety system protocols, subsea equipment, flowlines and risers, and control systems. Figure 3.2 illustrates some of these structures along with their corresponding recommended practices.

The main advantage of analyzing these structures as a system is the possibility to examine the parts, components, and equipment through their interdependencies across the entire project lifecycle, from conception to decommissioning.

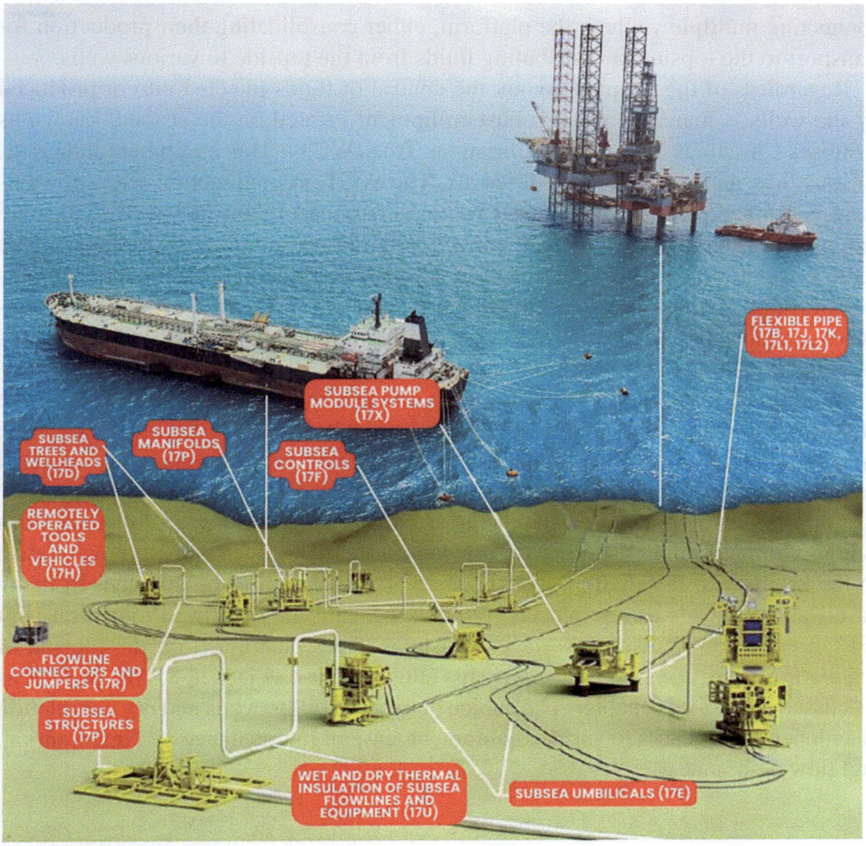

Fig. 3.2 Offshore structures along with their corresponding recommended practices

This approach also addresses issues related to technology qualification, engineering, inspection, training, and reliability.

Next, some of this equipment will be described, providing an overview of their primary functions, composition, and characteristics. Understanding these topics is essential for offshore operations engineering, as operations frequently involve wells, pipelines, offshore installations, support vessels, and service vessels.

3.1.1 Subsea Christmas Tree

A typical subsea layout in oil field development involves isolated subsea wells (satellite wells) connected to a production platform with dedicated production pipelines and control umbilicals for each well. Alternatively, subsea manifolds can be utilized, connecting multiple wells to the platform, either consolidating their production for transport to the topside or distributing fluids from the topside to various wells.

Regardless of the adopted layout, the control of fluids injected into or produced by the wells is managed using robust equipment located on the seabed, known as a Subsea Christmas Tree or Wet Christmas Tree (WCT). This equipment integrates various flow and pressure control valves. The WCT, in conjunction with the wellhead equipment, ensures that fluids remain contained within the well, withstand the internal pressures, and prevent contact with the surrounding environment. This setup facilitates the safe production of hydrocarbons, directing them to the flowlines that transport the production to the offshore installation. There are numerous references available that discuss the operation, configurations, positioning, and challenges associated with subsea Christmas trees [8, 30].

In addition to the flowlines, umbilicals connect the Wet Christmas Tree to the platform, allowing remote control of its valves. Alternatively, these valves can be operated directly on the WCT, but in deep waters, this requires a remotely operated vehicle (ROV) due to the limitations of human diving.

The field development decision using satellite wells and Subsea Christmas Trees is highly influenced by the reservoir drainage strategy, wellhead positioning relative to the production installation, and the type of installation to be adopted. In many cases, all wellheads can be placed vertically beneath the Floating Production Unit (FPU), enabling the use of Surface Christmas Trees positioned on the topside of the platform (dry completion wells). This allows the use of jacket, SPAR, or TLP, as discussed later in this chapter. The choice between satellite wells and dry completion also depends on access to surface equipment, ease of maintenance and intervention, and flow assurance.

3.1.2 Manifolds

As mentioned, one strategy for field development is to use manifolds to collect production from multiple satellite wells and send it to the production platform. This layout reduces investment in flowlines and umbilicals by minimizing the total length of lines needed, particularly when satellite wells are far from the production platform but close to each other. However, the downside of this approach is that the failure of a single subsea component can compromise the operation of several connected wells.

The manifolds enable: (i) individual control of production/injection from each well using choke valves; (ii) collection of production from all wells and its transfer to the platform via a single production flowline, while simultaneously injecting fluids into injector wells through an injection flowline; and (iii) well testing operations for any well connected to the manifold.

3.1.3 Flowlines and Umbilicals

As previously mentioned, the interconnection of Subsea Christmas Trees, manifolds, and offshore installations is achieved through pipelines known as flowlines. The length of these flowlines depends on the distance between the connected equipment and typically ranges from 3 to 8 km but can extend beyond 20 km for oil transport in some cases. In a standard scenario with satellite wells and manifolds linked to the production platform, it can generally be expected at least the following flowlines:

- **Production/Injection Flowline**: Allows the flow of the primary fluid intended for the well, whether it be oil production or water/gas injection.
- **Service Flowline**: Designed to pump a wide range of fluids from the FPU into the wells or to circulate them through the production flowline. This includes circulating diesel to prevent hydrates during well startup (after extended shutdowns), circulating gas to run a pig through the production line for paraffin wax removal, or injecting fluids into the well (bullheading[1]) for necessary commissioning.
- **Test Flowlines**: Present when using manifolds, it allows the selective direction of production from a single well to the FPU, enabling its fluids to be routed to the topside test separator. Meanwhile, the other wells connected to the manifold continue producing normally through the production flowline.

Flowlines can be constructed from rigid metallic pipes, typically made of carbon steel or corrosion-resistant alloys (CRA), such as duplex and super duplex steels, for environments with corrosive gases (e.g., CO_2 and H_2S) [18, 26]. As cited by Smith and Celant [27], "the selection of a material for a pipeline or flowline requires a compromise between optimum corrosion resistance, required mechanical properties, fabricability, availability and cost". The final choice may not always meet all requirements, but it must always consider the project's lifespan.

Flexible flowlines, which are quite common in offshore environments, are composed of a series of concentric layers, each serving a specific purpose [13, 28]. API RP 17B "provides guidelines for the design, analysis, manufacture, testing, installation, and operation of flexible pipes and flexible pipe systems for onshore, subsea, and marine applications" [5, p. 1], and according to this recommended practice, a flexible pipe "generally combines low bending stiffness with axial tensile and torsional stiffness, achieved by a composite pipe wall construction. The two basic components are helical reinforcement layers and polymer sealing layers, which allow a much smaller radius of curvature than for a steel pipe with the same pressure capacity" [5, p. 18].

[1] During petroleum production, bullheading involves pumping fluids into the well for a specific objective, without returning them to the surface. Precise pressure control is essential during these operations to avoid damaging pipelines, wellheads, and subsurface equipment. Unlike injection wells, bullheading does not aim to enhance oil recovery, as secondary recovery methods do. In drilling or workover operations, bullheading is typically considered a well control technique by pushing the unwanted fluids back into the reservoir.

The characteristics of these pipelines (diameter, composition, stiffness, among others) depend on the needs of the field development and must consider: the nature of the fluids to be produced (oil, water, or gas); maximum working pressures; possibility of hydrate, paraffin, or scale formation; seabed characteristics (topography, composition, and stability); existence of multiphase flow and flow pattern; presence of contaminants (H_2S and CO_2, for example).

Finally, the connection between a flexible pipe and a rigid pipe, or even between these and other structures, is possible through specific connectors, as can be seen in the Decommissioning Program for Maritime Installations of FPSO P-32 [33].

Umbilicals are an assembly of tubes and cables grouped into a flexible line with specific functions, such as:

- **Hydraulic Conduits**: Designed to transmit hydraulic pressure from a Hydraulic Power Unit (HPU) at topside for controlling remotely operated valves found in subsea equipment (seabed Christmas tree and manifolds) or in subsurface, such as DHSV or SSSV valves.
- **Chemical Injection Conduits**: Allows the injection of chemicals into the fluid streams at Christmas tree, wellhead, or well annulus. These chemicals can be injected continuously during system operation, such as the injection of demulsifiers, H_2S scavengers, and scale inhibitors, or in specific situations, such as the injection of alcohols (monoethylene glycol or triethylene glycol) during hydrate inhibition operations in the Subsea Christmas Tree.
- **Electrical Power and Signal Cables**: Designed to transmit digital signals from well variables for remote monitoring both on the offshore installation and on the onshore integrated monitoring centers. Common installations include pressure and temperature transducers (PTT) on the Subsea Christmas Tree to provide data on the pressure and temperature on the stream fluid. Additionally, pressure transducers (PT) deliver pressure readings at the Christmas tree point connected to the well annulus, and permanent downhole gauges (PDG) enable continuous monitoring of pressure near the reservoir, aiding in more effective reservoir management [10, 14]. This sort of cables can also provide power to seabed or subsurface pumps, such as electrical submersible pump (ESP) systems.
- **Fiber Optic Cables**: Used for remote monitoring of variables, fiber optic cables offer significant advantages over electrical cables. They enable the transmission of a larger volume of information at high speeds and are immune to magnetic interference, such as that generated by submersible electric pumps [12].

These conduits and cables are bundled together to form a single cylindrical unit, encased in multiple outer layers that provide mechanical strength and protection against abrasion and environmental conditions. The quantity and types of cables or conduits incorporated into the umbilical depend on the project requirements. Figure 3.3 illustrates a configuration of an umbilical with its various components.

Another type of umbilical that includes production and injection lines is the Integrated Pipeline Umbilical (IPU). According to API 17TR13, it "consists of a combination of one or more production and/or injection lines and/or various service lines,

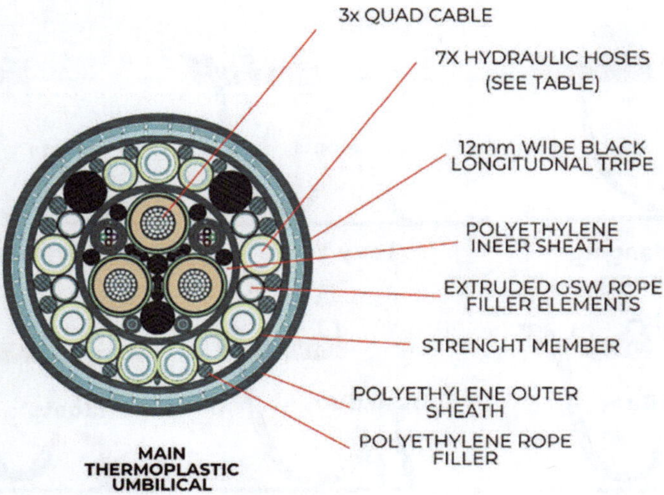

Fig. 3.3 Configuration of an umbilical with its various components

hydraulic lines, electrical and/or fiber optic cables, etc. An IPU differs from a traditional multicore umbilical in that it incorporates a relatively large-bore service or production line" [3, p. 61].

3.1.4 Risers

Risers are referred to as the portion of the flowlines between the Floating Production Unit (FPU) and the point where the pipeline touches the seabed (TDP—Touch Down Point). They are also called production risers (as they are connected to production platforms) and dynamic risers when the production platform is floating. Their composition does not differ significantly from flowlines, with flexible risers being commonly used in floating offshore installations, as they must be capable of absorbing the movements caused by the floating installation's motion.

When risers are connected to the platform from the seabed, they typically form a free-hanging catenary shape, which is common in various projects [24]. However, in deep and ultra-deep water projects, the weight of flowlines from multiple wells connected to the platform can be substantial. In such cases, alternative configurations can be employed using floats or buoys attached to the risers in different layouts (lazy wave, steep wave, Lazy-S, Steep-S, and pliant wave). These configurations offer an added benefit of absorbing vessel and wave motions, reducing loads on the TDP due to platform movement, preventing the displacement of this point along the pipeline, and consequently mitigating potential abrasion wear. Figure 3.4 shows this flexible riser layouts.

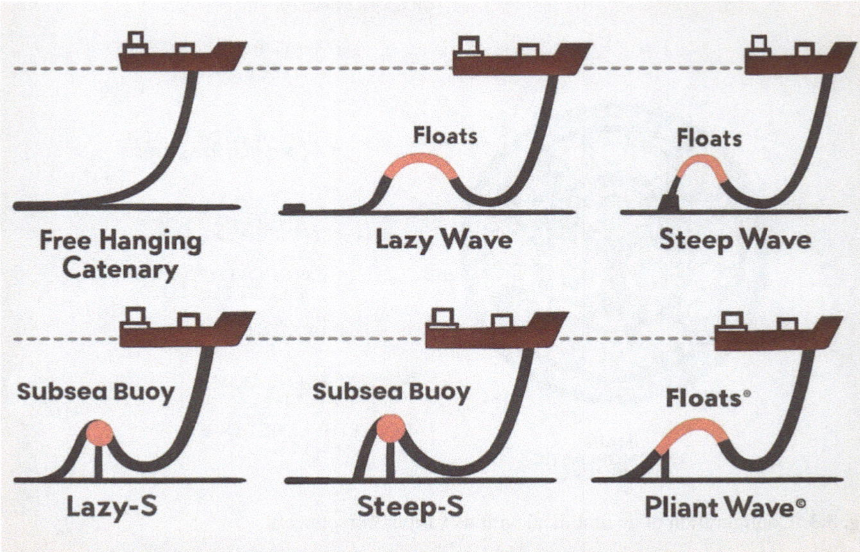

Fig. 3.4 Flexible riser layouts

3.1.5 Other Subsea Equipment

Advancements in subsea systems include equipment that, in addition to controlling flow, also perform some processing functions, such as phase separation of fluids (biphasic or triphasic), injection of produced water or seawater, and pressure boosting of fluids through single-phase or multiphase pumping. These structures will not be detailed in this work as they are specific to each type of field.

Information about all these components of subsea systems can be found from the main suppliers of flexible pipes (NOV, Baker Hughes, and TechnipFMC), umbilicals (Oceaneering and Prysmian), and equipment (Aker, Onesubsea, Baker Hughes, and TechnipFMC).

3.2 Offshore Production Installations

Offshore production installations are responsible for collecting fluids from wells, processing petroleum, and enabling the transport of these fluids to shore or storing them for later transfer to tankers.

Offshore production installations are commonly classified based on water depth and buoyancy. In this context, there are fixed installations on the seabed, which are more suitable for shallow waters (such as fixed platforms and compliant towers), and floating installations used in deep water (TLP, SPAR, FPSO, and semi-submersibles)

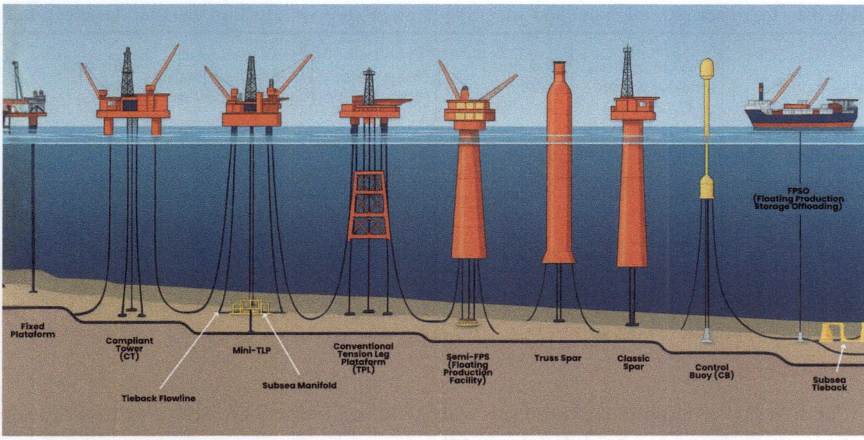

Fig. 3.5 Types of offshore production installations

and ultra-deep water (FPSOs and semi-submersibles). Figure 3.5 illustrates some of these installations.

Regarding floating production units (FPUs), another classification criteria is based on the environmental and metoceanographics conditions where the installation will operate. Units more suitable for milder and more stable environmental conditions throughout the year include FPSOs and SSs, while TLPs and SPARs are better suited for harsher conditions with greater annual variation. Understanding the environment where the platform is located is crucial for assessing its effects on stability and buoyancy,[2] because it directly impacts its design due to the stresses it will endure. For example, structures designed for production in the Gulf of Mexico must account for hurricane loads during both the design and operation phases [7], whereas those intended for Southeast Brazil, Southeast Asia, or West Africa face calmer environments with fewer annual variations.

After an evaluation that considers the water depth, metoceanographic conditions, and the number of wells to be exploited, Ronalds [25] presents the main applications for at least eight types of offshore production installations. Table 3.1 summarizes some conclusions of this work, where a few observations can be made:

- For deep water, the number of wells dictates the choice of the production unit in the case of dry completions (TLP or SPAR). This is tied to the tension forces required in the risers, the buoyancy capacity of the hull, and ultimately, the tension in the mooring tendons. Consequently, there is a range of applications from mini TLPs to SPARs and full TLPs.
- FPSO units for shallow waters are typically linked to short-term projects with few wells, where constructing a fixed platform might not be justified.

[2] More informations about this subject can be found in Chap. 5.

Table 3.1 Influence of well count and water depth on offshore production installation selection

Well count	Shallow to mod shallow water (< 70–300 m)	Mod deep water (300–500 m)	Deep water (500–1350 m)	Very deep water (1350–1800 m)	Ultra-deep water (> 1800 m)
Very high (> 36)		Compliant tower			
High (20–36)	Fixed platforms		TLP		
Intermediate (9–20)			SPAR		FPSO
Mod low (5–9)			SPAR/mini TLP		

Adapted from Ronalds [25]
Shadow area represents water depths were FPSOs and SSs are suited to operate

- For lower well counts and shallow water, subsea tiebacks to shore or to a remote host can be used for field development.

Additional factors influencing the selection of a production unit include the field development strategy, which is intricately linked to reservoir drainage and the optimal utilization of produced fluids. This strategy will also require evaluating whether a drilling/workover rig should be present on the production unit and the type of completion to be adopted (dry or wet).

It is important to note that it was mentioned only structures involved in oil production, thereby excluding offshore installations with different purposes, such as mobile offshore drilling units (MODU), offshore wind turbines, and liquefied natural gas carrier vessels. The following sections will cover the most common types of offshore production installations, highlighting their key characteristics and applications. However, this text does not intend to provide an exhaustive discussion. For a deeper exploration of the topic, other research sources can be consulted [11].

3.2.1 Fixed Platforms

Fixed platforms are stationary units secured to the seabed by "legs" made of concrete or steel. These legs are usually supported on piles driven into the ground, but they can also be laid down direct to the seafloor depending on total weight and expected environmental conditions. This design provides minimal movements due to currents and winds, which is the main advantage of this type of platform. The casing conductors of the well can be extended from the seafloor to the installation deck, allowing the wellhead to be placed there and hence, enabling the use of dry trees.

Economic feasibility linked to leg's length make conventional fixed platforms suited for shallow waters, typically near 200–250 m, as can be found at Gulf of Mexico (GOM). Over that depth, Compliant Towers are more appropriate due to the flexibility of their legs and better response to oceans forces. One of the tallest compliant tower in the world is the Petronius Oil Platform, located in the Gulf of

México, 210 km southeast New Orleans, operated by Chevron and anchored in water depths of 535 m.

3.2.2 Semisubmersible Units (SS)

A typical design of a semisubmersible platform consists of a large deck supported by four or six vertical columns connected at the bottom to two horizontal pontoons. Usually, after construction, the hull integration is carried out on the shipyard and the semisubmersible is dry towed to the location to remain permanently stationed by anchors with mooring systems (chains, wires or polyester ropes). The pontoons and the columns are partially filled with water to ballast control, providing weight and buoyancy.

Semisubmersibles are suited for deep water and harsh environments due to their high stability, allowing the use of wet trees and flexible risers. Oil and gas production must be pumped to pipelines due the lack of storage capability.

3.2.3 Floating Processing, Storage and Offloading Unit (FPSO) Type Units

FPSOs (Floating Processing, Storage and Offloading) are large vessels mainly used to develop oilfields located in deep and ultra-deep water and/or not connected to pipeline systems capable of delivering oil production to shore.

The oil production from subsea wells flows through a processing facility on the FPSO topside, where several devices such as heaters, scrubbers, and separator vessels are installed to separate oil, water and gas. Once this initial separation is completed, the oil is stored in large capacity cargo tanks, where it remains until being offloaded to a tanker by cargo pumps and through a flexible hose. The water can be reinjected into the reservoir through injection wells or discharged directly into the ocean after removal of oily phase, typically below 25–15 ppm depending on regional environmental regulations.

The destination of produced gas depends on a series of circumstances. Typically, after dewatering and contaminant-removal treatment (CO_2 and H_2S), a portion of the gas is used as fuel in gas turbines. Another portion may be used for artificial lift in production wells (gas lift) or injected into the reservoir for enhanced oil recovery purposes. Finally, if a subsea pipeline is available, some of the produced gas can be exported to shore. All these uses require a gas treatment and compression facility, typically equipped with scrubbers, filters, multi-stage compressors, and dehydration units (glycol contactors or molecular sieves).

There are two main alternatives for building an FPSO. The first one is to convert an oil tanker, i.e., a Very Large Crude Carrier (VLCC), to meet the requirements of a

FPSO, such as adaptations on the pump room, tanks (cargo, ballast and slop), power generation, accommodation, fire extinction and mooring system, and also installation of all the oil process facility, gas treatment and compression, water/gas injection and offloading system.

The other alternative is to construct a newbuild vessel using the principles of a tanker hull design as a starting point [21]. This way, it can be fitted to meet most desirable project conditions and up-to-date regulations, such as double hull and bottom requirements of MARPOL convention.[3] The main advantage of conversion is that the FPSO can be delivered in less time than a newbuild, which may be an important driver depending on the field development strategy. Also, conversion is usually less expensive than a newbuild, but useful life must be considered.

As the tendency of oil production in Brazil is to establish itself in increasingly deeper water, most of the new projects are for Stationary Floating Units such as the SS and FPSO. Hence, over the last decade Brazil has become one of the leaders on FPSO installations, and Petrobras is the company responsible for this achievement, with more than two dozen new FPSO operating on the development of the pre-salt reservoirs in Santos Basin and revitalization of mature oilfields in Campos Basin.

3.2.4 Single Point Anchor Reservoir—SPAR

A SPAR consists of a deck supported by a large deep-draft cylindrical hull anchored to the seabed by vertical mooring lines or tendons. This design ensures that the center of gravity is lower than the center of buoyancy, providing stability and enabling it to withstand harsh environmental conditions, such as strong winds and currents. Depending on the shape and construction of the hull, a SPAR can be classified as:

- Classic or conventional SPAR: a single huge cylindrical hull that floats vertically in the water providing stability.
- Truss SPAR: the hull is divided into three main components. At the top, a short cylindrical hull provides primary buoyancy. In the middle, a tubular truss structure connects the upper and lower modules. At the bottom, a soft tank serves primarily for ballast and motion absorption.
- Cell SPAR: the hull consists of a cylindrical module in the center surrounded by multiple independent cylindrical cells with diameters ranging from 6 to 15 m. As the truss SPAR, the soft tank is in the lower section.

This type of unit offers the advantage of enabling the use of dry Christmas trees. Additionally, sections of the hull can be used for ballast control and storage.

---

[3] To prevent oil pollution in an event of collision, MARPOL requires that the entire cargo tank length shall be protected by ballast tanks or spaces other than tanks that carry oil. This applies to wing tanks or spaces and double bottom or spaces. See IMO [32] for more information.

Table 3.2 Main advantages and disadvantages of each type of offshore installation

Offshore installation	Advantage	Disadvantage
Fixed platform	Independent from environmental conditions Allows installation of drilling and completion rigs Dry Christmas tree	Limited operational water depth Dependent of seabed stability to installation
Semi-submersible	Large deck area for process plant equipment Adequate to harsh environment	No storage capability Wet Christmas tree High maintenance costs
FPSO	Storage capability Large deck area for process plant equipment Wide range of water depth applicability Conversion of tankers may lower cost and early production	Wet Christmas tree Poor stability in harsh environments
TLP	Dry Christmas tree Mobility and reusability Adequate to harsh environment	No storage capability Tendon or tension legs fatigue High initial cost
SPAR	Dry Christmas tree Adequate to harsh environment (good stability) Low maintenance cost	Difficulty and complex installation process Little storage capability

3.2.5 Tension Leg Platforms (TLP)

Tension Leg Platforms (TLPs) are vertical structures anchored to the seabed by tethers or tendons. These tendons have low elasticity, which reduces the impact of heave motion on the platform, thereby allowing the use of dry Christmas trees. However, the cost of tendons and risers limits the installation depth of TLPs, making them unsuitable for ultra-deep water applications.

Table 3.2 shows some advantages and disadvantages of each type of offshore installation unit presented in this chapter.

3.3 Simultaneous Operations (SIMOPS)

There are several definitions of Simultaneous Operations (SIMOPS) in the petroleum industry, depending on the level of activities being evaluated. Basically, simultaneous operations are situations where two or more activities are performed at the same time, which can lead to additional risks to one another. During the routine operation of a platform, the increased risk typically arises from the proximity of various activities. For example, performing a naked flame work near a pipe painting operation or depressurizing a scrubber can significantly elevate the risk of these tasks.

On the other hand, the risk level can be altered when an activity, even conducted far from another, affects the mitigation of a potential incident or increases its severity. For instance, this can occur when the firefighting system is unavailable due to maintenance while the process plant is being started up.

When we analyze SIMOPS beyond the activity level and start evaluating operations involving offshore production facilities, several examples of simultaneous operations can be found, such as: operations involving a production unit and support vessels; the operation of an offshore production unit linked to an accommodation vessel; drilling, workover and construction activities occurring simultaneously with oil production on the platform; and ROV activities on wellheads with remote assistance from the production platform.

In this context, there have been serious personnel, maritime, and process safety accidents in the past, resulting in fires, explosions, and collisions, which have led to multiple fatalities. Hence, a multidisciplinary analysis and proper risk management of SIMOPS "can enhance oil and gas companies' operations and reputation through: improving contractors' HSE performance (...); improving the relation with contractors and creating an open and transparent environment for discussing all operational and HSE issues (...); setting practical measures to reduce risks to ALARP level and communicating HSE requirements to all contractors" [6, p. 4].

A common approach to addressing these issues is to establish, prior to the activities, a matrix that defines the conditions under which SIMOPS can occur, or even indicates when they cannot occur. For instance, a restriction might specify that an offloading operation with a tanker can only be performed simultaneously with an ROV inspection on risers if there is sufficient distance between the vessels involved in the SIMOP.

As an example of SIMOPS analysis, Ologe et al. [23, p. 1] presents a case study about some complex drilling, production, and heavy construction operations performed simultaneously in a upgrade project at Nigeria, with some uncharacteristically elements such as: "more than 3 years non-stop activities, more than 6 million man-hours, more than 4000 people working together (...), involvement of consortium of 3 contractors and, more than 15 major subcontractors"; large number of thermal engines (mobile and stationary) and other ignition sources in hazardous zones and so on. Additionally, Al Hammadi et al. [2] described the challenges faced during the first gas SIMOPS operation in Abu Dhabi offshore, where drilling operations were performed simultaneously with gas injection on the same Well Head Tower. There are plenty of other applications considering fields and complex projects [9].

On a daily basis, SIMOPS evaluations can be conducted during the planning phase of the Permit to Work (PTW) process. During this phase, all activities expected to occur on the site are assessed for potential interference with one another. The PTW system must be an integral part of the company's Safety Management System, with the objective of identifying potential hazards during job performance and recommending measures to avoid these hazards and minimize the likelihood of an accident.

Consequently, a PTW serves as formal authorization to carry out a job and must be well understood by the workers on the front line.

A PTW system is crucial to operational safety, and several incidents in the Oil and Gas industry have occurred due to deficiencies in the PTW System. As stated by Lord Cullen, on the public inquiry into the Piper Alpha incident: "A PTW System is an essential part of a procedure to ensure that the work is done safely. Safety in this context means the safety not only of those carrying out the work, but also of those who may be affected by the carrying out of that work. An examination of the system as it prevailed on Piper for a substantial period up the time of the disaster raised several general questions" [29, p. 191]. Therefore, a thorough understanding of its framework, implementation, and practice is strongly recommended for all positions involved in offshore engineering operations.

Regarding to SIMOPS with multiple vessels, Kallaway [17, p. 2] states that "SIMOPS coordination is a collaborative, risk-based mitigation process using a formal written exchange of information between concerned parties or vessels. The goal of Managing SIMOPS is to reduce the potential clash of vessel activities which could bring about an undesired event in normal or emergency situations. Whenever multiple vessels work in close proximity, complexity increases, and the risk of an incident rises significantly. SIMOPS Management is an Industry Best Practice for conducting safe and efficient vessel coordination when responding to a marine-based Source Control Event (SCE)".

References

1. B.S. Aadnoy, I. Cooper, S.Z. Miska et al., *Advanced Drilling and Well Technology* (Society of Petroleum Engineers, Richardson, 2009). https://doi.org/10.2118/9781555631451
2. M.I. Al Hammadi, A. Scheed, H. Alsabri et al., First ever gas SIMOPS implementation in Abu Dhabi offshore: a success story, in *Abu Dhabi International Petroleum Exhibition & Conference*, Abu Dhabi, 15–18 Nov 2021. https://doi.org/10.2118/207359-MS
3. American Petroleum Institute, *Technical Report 17TR13: General Overview of Subsea Production Systems* (American Petroleum Institute, Washington, DC, 2016)
4. American Petroleum Institute, *API RP 17A: Design and Operation of Subsea Production Systems—General Requirements and Recommendations* (American Petroleum Institute, Washington, DC, 2022)
5. American Petroleum Institute, *API RP17B: Recommended Practice for Flexible Pipe* (American Petroleum Institute, Washington, DC, 2024)
6. Z.M. Ataya, New technique for addressing SIMOPS challenges during installation of new offshore platform, in *The SPE Saudi Arabia Section Technical Symposium*, Al-Khobar, 10–12 May 2008. https://doi.org/10.2118/120796-MS
7. S.W. Balint, D. Orange, Panel discussion: future of the Gulf of Mexico after Katrina and Rita, in *The Offshore Technology Conference*, Houston, 1–4 May 2006
8. R.V. Beppler, A. Rabello, E.J.J. Coelho et al., Wet Christmas tree and control system standardization and interchangeability—more than 10 years of lessons learned, in *The Offshore Technology Conference*, Houston, 6–9 May 2024. https://doi.org/10.4043/35189-MS
9. A. Bird, J. Lumsden, J. Paterson et al., Case study: first multistage fracturing SIMOPS with concurrent drilling operations in Clair Ridge Field, North Sea, in *The SPE International*

Hydraulic Fracturing Technology Conference and Exhibition, Muscat, 12–14 Sept 2023. https://doi.org/10.2118/215632-MS
10. C.J.H. Chris, N.S. Nordin, M.A. Nuriyadi et al., Permanent downhole gauge: a need or a luxury?, in *The Asia Pacific Oil and Gas Conference & Exhibition*, Jakarta, 4–6 Aug 2009. https://doi.org/10.2118/122604-MS
11. E.C. Edwards, A. Holcombe, S. Brown et al., Evolution of floating offshore wind platforms: a review of at-sea devices. Renew. Sustain. Energy Rev. **183**, 113416 (2023). https://doi.org/10.1016/j.rser.2023.113416
12. M. Etuhoko, M. Viti, S. Zmeyevskiy et al., Deployment of downhole fibre optical sensing technology in the Karachaganak Field, Kazakhstan, in *The SPE Annual Caspian Technical Conference & Exhibition*, Baku, 4–6 Nov 2015. https://doi.org/10.2118/177387-MS
13. C. Fernández, C. López, Use of flexible flowlines for offshore conditions in the Lake Maracaibo oil/gas fields, in *The CORROSION 2014*, San Antonio, 9–13 Mar 2014
14. H.M. Frota, W. Destro, Avaliação da confiabilidade de sistemas de monitoramento permanente de fundo de poços (PDG/TPT) no campo de Marlim, in *Rio Oil and Gás* (2024)
15. J. He, H. Zhang, Z. Renchuan, Study on dynamic positioning capacity analysis method considering wind shielding effect of offshore structure. J. Ship Res. **66**(4), 315–325 (2022). https://doi.org/10.5957/JOSR.12200062
16. International Maritime Organization, *MSC.1/Circular.1580—Guidelines for Vessels and Units with Dynamic Positioning (DP) Systems* (2017). https://www.register-iri.com/wp-content/uploads/MSC.1-Circ.1580.pdf. Accessed 20 Apr 2025
17. K. Kallaway, Subsea well response project—simultaneous operations (SIMOPS) multi vessel operations—guidance for marine based response to subsea well incident, in *The Offshore Technology Conference*, Houston, 4–7 May 2015. https://doi.org/10.4043/25784-MS
18. A. Kharusi, J.R. Saithala, A.O. Macedo et al., Flowlines material performance and challenges for oil & gas applications, in *The CORROSION 2020*, Online Event, 14–18 June 2020
19. L.W. Lake, R.F. Mitchell, *Drilling Engineering* (Society of Petroleum Engineers, Richardson, 2006). https://doi.org/10.2118/9781555631147
20. K.-T. Ma, Y. Luo, T. Kwan et al., *Mooring System Engineering for Offshore Structures* (Gulf Professional Publishing, Cambridge, 2019). https://doi.org/10.1016/B978-0-12-818551-3.00009-0
21. A. Mather, *FPSO Handbook* (Whiterby Seamanship International Ltd., Fort Lauderdale, 2009)
22. R.F. Mitchell, S.Z. Miska, *Fundamentals of Drilling Engineering* (Society of Petroleum Engineers, Richardson, 2011). https://doi.org/10.2118/9781555632076
23. J. Ologe, K. Akiode, T. Jinkoji et al., Managing health, safety and environment in a complex simultaneous operations (SIMOPS) within the context of an integrated projects: a case study, *The SPE International Conference on Health, Safety, and Environment*, Long Beach, 17–19 Mar 2014. https://doi.org/10.2118/168428-MS
24. F.O. Orlando, Análise de Instalação de risers Lazy Wave em águas ultraprofundas. Dissertação, Universidade Federal do Rio de Janeiro, 2019
25. B.F. Ronalds, Applicability ranges for offshore oil and gas production facilities. Mar. Struct. **18**(3), 251–263 (2005). https://doi.org/10.1016/j.marstruc.2005.06.001
26. J.R. Saithala, T. Illson, I. Thompson et al., Corrosion management of duplex stainless steel gas production flowlines, in *The CORROSION 2014*, San Antonio, 9–13 Mar 2014
27. L.M. Smith, M. Celant, Life cycle costing—are duplex stainless steel pipelines the cost-effective choice?, in *The Offshore Technology Conference*, Houston, 1–4 May 1995. https://doi.org/10.4043/7789-MS
28. R.L. Tanaka, R.G. Morini, C.A. Godinho et al., Qualification of flexible flowlines: lateral instability testing, in *The Offshore Technology Conference*, Houston, 5–8 May 2014. https://doi.org/10.4043/25126-MS
29. The Hon Lord Cullen, *Public Inquiry into the Piper Alpha Disaster* (HMSO, London, 1990). https://www.hse.gov.uk/offshore/assets/docs/piper-alpha-public-inquiry-volume1.pdf
30. M. Yamamoto, K. Miura, The application and functionalities of a wet Christmas tree applied in Santos Basin pre-salt cluster, in *The 2017 Annual Autumn Meeting*, The Japan Society of Naval Architects and Ocean Engineers, Tokyo, 27–28 Nov 2017

Bibliography

1. International Maritime Organization, *Annex I of MARPOL 73/78 Regulations for the Prevention of Pollution by Oil* (2024). https://www.imorules.com/MARPOL_ANNI.html. Accessed 20 Apr 2025
2. Petrobras, *Programa de Descomissionamento de Instalações Marítimas (PID) do FPSO P-32* (2020). https://www.gov.br/anp/pt-br/assuntos/exploracao-e-producao-de-oleo-e-gas/seguranca-operacional/arq/ppdi/pdi-fpso-p-32.pdf/@@download/file. Accessed 16 Mar 2024
3. R.B. Veloso, A.R. Pereira, M.R.R. Augustus, Subsea X-Mas tree standardization benefits in Brazil scenario, in *The Offshore Technology Conference*, Brasil, Rio de Janeiro, 29–31 Oct 2019. https://doi.org/10.4043/29872-MS

Chapter 4
Basic Notions of Naval Structure

Abstract The chapter "Basic Notions of Naval Structure" introduces the fundamental principles that govern the structural behavior of ships and offshore platforms, emphasizing the types of loads acting on these constructions and the key methods used in their structural analysis. It classifies loads into gravitational, buoyancy, environmental, and operational, explaining their origins, characteristics, and relevance for design and safety. The chapter also presents essential concepts of naval structural calculation, including light ship displacement, hull girder theory, bending moment, and shear force, which describe how ships respond to internal and external stresses. Furthermore, it discusses the global structural analysis of vessels using simplified "box girder" models to evaluate longitudinal strength under conditions such as hogging and sagging. Through these foundational topics, the chapter provides a clear understanding of how engineers assess the strength, stability, and durability of naval and offshore structures throughout their service life.

4.1 Loads Acting on Offshore Structures

Throughout their useful lives, offshore constructions, such as fixed platforms, floating platforms, ships and vessels, experience several loads throughout their lifespan. These loads can be categorized into four basic types: gravitational, buoyancy, environmental, and operational. It is crucial to comprehend these loads for the design, construction, and operation of offshore structures.

4.1.1 Gravitational Loads

Dead loads, commonly referred to as gravitational loads, refer to the static or non-moving weight of the ship and its permanent fixtures. These loads include the weight of the ship's hull, superstructure, machinery, equipment, fuel, ballast, and other permanently installed components. Dead loads also encompass any fluids (such as

M. Vinicius da Silva Neves et al., *Offshore Operations Engineering*,
SpringerBriefs in Applied Sciences and Technology,
https://doi.org/10.1007/978-3-032-04715-1_4

oil, water, or other liquids) that are permanently stored within the ship. Due to gravity, these loads act vertically downward and continuously. By accurately determining and accounting for dead loads, naval architects can ensure that the ship maintains proper balance and stability during its operation life.

4.1.2 Buoyancy Loads

The forces that the structure is subjected to, because of being submerged in water, are known as buoyancy loads. The buoyant force on an object is equal to the weight of the fluid it is dislodging, according to Archimedes' principle. To maintain stability for floating constructions, the buoyancy load and the gravity load must be equal.

4.1.3 Environmental Loads

Environmental loads are forces exerted on the structure by environmental factors. These include:

1. **Wind Loads**: The force exerted by wind on the structure. Wind loads can cause both drag and lift forces and can vary greatly depending on the wind speed and direction.
2. **Wave Loads**: The force exerted by waves on the structure. Wave loads can cause significant stress on offshore structures, particularly during storm conditions.
3. **Current Loads**: The force exerted by ocean currents on the structure. Current loads can cause drag forces and can affect the stability of the structure.
4. **Ice Loads**: In colder climates, the force exerted by ice on the structure can be significant. Ice loads can cause crushing forces and can also lead to increased drag.

4.1.4 Operational Loads

Operational loads are the forces that emerge during the operational phase of a structure. They encompass the following:

1. **Live Loads**: These are transient loads that arise during the operation of the structure, such as the weight of personnel, temporary equipment, and consumables.
2. **Dynamic Loads**: These forces stem from the movement of machinery or equipment on the structure. Dynamic loads can induce vibrations that must be considered in the structural design.

3. **Impact Loads**: These forces are generated by activities such as drilling or helicopter landings, for example. These loads can lead to significant stress on the structure and must be carefully considered during the design phase. Taking into account impact loads is crucial to ensure the structural integrity and safety of the designed structure.
4. **Cargo Loads**: Cargo loads refer to the forces resulting from the weight of fluids stored, transferred, or processed during the operation of offshore production units. These loads include the mass of produced fluids such as oil, produced water, and injection water, as well as ballast water used to control the vessel's stability and draft. The distribution and variation of these loads directly affect the structural behavior and must be accurately considered in both design and operational assessments.

   For example, significant cargo loads may arise from the temporary storage of: Produced Water: heavier than oil, can accumulate in large volumes in dedicated tanks or separators;

   Crude Oil: typically stored prior to offloading, contributing to substantial weight;

   Ballast Water: dynamically managed to maintain trim and stability, often shifted between tanks during operations.

   Proper evaluation of cargo loads is essential for assessing global hull strength, local tank structure design, and maintaining structural integrity throughout the operational lifecycle.

It is essential for the design and operation of offshore structures to comprehend the many sorts of loads that they are subjected to. Engineers must consider all these loads to guarantee the effectiveness and safety of offshore buildings.

4.2 Naval Structural Calculation

A crucial component of ship design and maintenance is naval structural calculation, which makes sure the ship can endure a variety of forces and stresses while in use. The fundamental ideas of naval structural calculation will be covered in this chapter, along with important concepts like the Hull Girder Theory, Bending Moment, Shear Force, and others.

4.2.1 Light Ship Displacement

The displacement of a vessel corresponds to the mass of water displaced by it when afloat and is usually expressed in tons. This value is obtained by multiplying the immersed volume of the vessel by the density of the water in which it navigates. In practice, displacement represents the mass of the vessel itself at a given moment.

As a measure of mass, displacement should not be confused with tonnage, which a measure of the cargo-carrying capacity of a ship, and is commonly used to assess fees on commercial shipping, and should also not be confused with deadweight, as deadweight (DWT) or tons deadweight is a measure of how much weight a ship can carry.

When simply referring to the displacement of a ship, it is understood to be referring to its maximum displacement.. The weight of a light ship, on the other hand, refers to the weight of the ship with all of its permanent fixtures and equipment on board, but without any cargo, fuel, water, ballast, stores, passengers, or crew.

A key factor in ship design and stability studies is the light ship displacement. It is used to figure out the ship's stability qualities and draught, or the amount of water required to float the ship.

Light Ship Displacement Calculation

The weights of all the structural, operational, and equipment components of the ship can be added up to determine the light ship displacement (D). The hull, superstructure, propulsion system, electrical system, navigational equipment, and other components fall under this category.

$$D = \text{component weights}$$

In practice, the weights of the individual components are often estimated based on empirical formulas or data from similar ships. The total light ship displacement is then adjusted as necessary during the ship's construction and sea trials to match the actual measured displacement.

Example: Calculating Light Ship Displacement

Let's consider a simple example. Suppose we have a ship with the following component weights:

- Hull: 5000 tons
- Superstructure: 1000 tons
- Propulsion system: 500 tons
- Electrical system: 200 tons
- Navigation equipment: 50 tons.

The light ship displacement is then:

$$D = 5000 + 1000 + 500 + 200 + 50 = 6750\,\text{tons}$$

So, the light ship displacement of the ship is 6750 tons.

4.2.2 Hull Girder Theory

A fundamental concept in naval architecture is the "Hull Girder Theory," which proposes to view a ship as a beam or girder that experiences various forces like waves-induced loads, cargo loads, buoyancy, and self-weight. This concept highlights the theoretical box girder formed by the transverse sections and continuous longitudinal structural members of a ship's hull, which provides resistance to hogging and sagging when it is floating.

To comprehend how a ship's structure responds to various forces, it is essential to understand the Hull Girder Theory. The key components of the hull girder include the hull plates, secondary girders, and main girders. Together, they work to ensure the structural integrity and stability of the vessel under different loading conditions.

The hull plates help distribute loads and stresses experienced by the ship across the structure, while the girders serve as the primary load-bearing elements within the hull, effectively transferring loads from the hull plates to the ship's structure and ensuring structural stability.

The fundamental equation for a beam's bending stress is:

$$\sigma = \frac{M * y}{I}$$

where:

- σ is the bending stress
- M is the bending moment at the section
- y is the distance from the neutral axis
- I is the moment of inertia of the section.

4.2.3 Bending Moment

The reaction experienced by a structural element when an external force or moment is applied, causing the element to bend, is known as the bending moment in a ship. A ship's greatest bending moment often occurs amidships, where the opposing forces of weight and buoyancy are most pronounced. For structural design and analysis, understanding the distribution of bending moments along the ship's length is essential.

The bending moment at a section due to a force F at a distance D from the section is given by:

$$M = F * D$$

4.2.4 Shear Force

Shear force refers to the internal forces generated within a ship's structure in response to applied external loads. Just like bending moments, shear forces are distributed along the length of the ship and are crucial to its structural integrity. By integrating the load distribution along the ship's length, one can determine the shear force distribution.

The shear force V at a section due to a force F is simply:

$$V = F$$

4.2.5 Other Relevant Topics

Other relevant topics in naval structural calculation include torsional moments, longitudinal strength, and structural response to wave loads. Torsional moments arise from forces causing a ship to twist around its longitudinal axis. Longitudinal strength is a measure of a ship's ability to withstand the bending moments and shear forces acting along its length. The structural response to wave loads involves understanding how a ship's structure responds to forces exerted by waves, which can cause bending, shear, and torsional stresses.

Torsional Moments: Torsional moments in a ship can be calculated by considering the forces that cause the ship to twist about its longitudinal axis. This typically involves the use of torsion theory, which includes equations for calculating shear stress and angle of twist in a shaft under a given torsional load.

The shear stress τ due to a torsional moment T in a circular shaft is given by:

$$\tau = \frac{T * r}{J}$$

where:

- τ is the shear stress
- T is the torsional moment
- r is the distance from the center of the shaft
- J is the polar moment of inertia of the section.

Longitudinal Strength: The longitudinal strength of a ship can be assessed by considering the ship's ability to resist bending and shear forces along its length. This involves calculating the bending stress and shear stress in the ship's structure and comparing these to allowable stress limits.

The bending stress σ due to a bending moment M is given by:

$$\sigma = \frac{M * y}{I}$$

where y is the distance from the neutral axis, and I is the moment of inertia of the section.

Structural Response to Wave Loads: The response of a ship's structure to wave loads can be calculated using wave load models and structural analysis methods. This involves calculating the wave-induced pressures on the ship's hull and then determining the resulting stresses and deflections in the ship's structure.

This is a complex topic that generally requires numerical methods for solution. The wave loads on a ship can be calculated using linear wave theory or other wave load models, and the resulting stresses and deflections can be calculated using the methods described above.

These are very simplified versions of the equations used in naval structural analysis, and many factors have been omitted for clarity. For a more detailed and accurate analysis, more complex methods and equations would need to be used, taking into account factors such as the distribution of material in the section, the effect of shear deformation, the dynamic nature of the loads, and the material properties of the structure.

4.2.6 *Simple Calculation Example*

A traditional and fundamental calculation in Naval Engineering is the global structural analysis of a vessel, aimed at verifying whether the vessel's structure can withstand all the loads it is subjected to. The most common methodology first involves simplifying the vessel to a "box girder" model, with a length equivalent to approximately three of the longest tank of the vessel, and only later elaborating more detailed calculations.

In this structural analysis strategy using a global model of three tanks, the box girder is subjected to anticipated operational loads, as well as environmental loads that occur with a frequency of one hundred years. The goal is to ensure that the vessel can withstand, without structural failure, the loads it will face throughout its service life, even if they are rare, like the centennial environmental occurrences.

Global analyses mainly focus on longitudinal bending and shear strength, which are the most critical stresses faced by marine structures. Torsional and transverse shear strength are occasionally analyzed after the longitudinal analysis, depending on the vessel, since longitudinal stresses are generally greater and often responsible for observed structural failures.

To illustrate the concept mentioned above, let's imagine a calculation that represents a highly simplified model, as real ship structures are much more complex and would depend on computational calculations. The aim here is to illustrate the concept of calculation, which would be based on the standards of Classification Societies and computational models using Finite Element Analysis.

Consider a solid box with thin thickness, with a length equivalent to approximately three times the length of the longest tank of the vessel. The thickness of the box represent the most commonly thicknesses of the hull deck, bottom and its portside and starboard. This box may or may not include vertical structures representing the presence of watertight longitudinal bulkheads.

Under this box, vertical loads related to buoyancy, the weight of the cargo, and the vessel's own weight, which includes all equipment and structures, are applied. Additionally, moment loads and shear forces caused by wave passage are considered. These loads are generally determined by the standards of classification societies and consider the vessel in bending in two directions: Sagging and Hogging (*).

(*) Sagging: This is the bending of the vessel that occurs when it is supported at the ends of the wave crests, causing the midship region to sag relative to the bow and stern.

(*) Hogging: This is the opposite of sagging, when the midship region is supported at the crest, while the bow and stern ends hang over the troughs on each side.

Using this methodology, it is possible to calculate approximate stress values based on classical calculations from the Structural Resistance discipline applied to slender beams. To do this, one must determine the area of the box girder's section and its Section Modulus. With the shear forces and moments in waves, it is possible to calculate the global bending moments and shear forces of the vessel, checking if they exceed the material's resistance limits, usually steel.

It is important to emphasize that this is a highly simplified example and that real ship structures are much more complex. In a real vessel, factors such as hull shape, material distribution, the effect of shear deformation, and the dynamic nature of loads must be considered. Additionally, the vessel is subject to loads related to specific operations, as well as environmental loads (wave loads, wind loads) and loads due to the vessel's acceleration in motion. For a more accurate analysis, more complex methods and software tools, such as computational finite element analyses, are typically used.

Bibliography

1. C.V. Amaechi, A. Reda, H.O. Butler et al., Review on fixed and floating offshore structures. Part II: sustainable design approaches and project management. J. Mar. Sci. Eng. **10**(7), 973. https://doi.org/10.3390/jmse10070973 (2022). This paper reviews the loads acting on offshore structures and discusses sustainable design approaches and project management
2. N.D.P. Barltrop, G.M. Mitchell, J.B. Attkins, *Fluid Loading on Fixed Offshore Structures* (HMSO, London, 1990). https://core.ac.uk/download/pdf/237397122.pdf. Accessed 21 Apr 2025. This document reviews the experimental work performed on wave particle kinematics and fluid loading in the context of the design of fixed offshore structures

3. M.A. El-Reedy, *Marine Structural Design Calculations* (Butterworth-Heinemann, Oxford, 2014). This book is a "quick look up guide", which includes both fps and SI units and is divided into categories such as Project Management for marine structures
4. O.M. Faltinsen, *Sea Loads on Ships and Offshore Structures* (Cambridge University Press, Cambridge, 1993). This book introduces the theory of the structural loading on ships and offshore structures caused by wind, waves, and currents
5. P.A. Frieze, *Offshore Structure Design and Construction* (PAFA Consulting Engineers, 2011). https://www.eolss.net/sample-chapters/c05/E6-177-OD-01.pdf. Accessed 18 Apr 2025. This document defines temporary and in-service design requirements, involving appropriate loads and combinations, as well as the demands for structural analysis and more
6. E.V. Lewis, *Principles of Naval Architecture Second Revision* (The Society of Naval Architects and Marine Engineers, Jersey City, 1988). https://ftp.idu.ac.id/wp-content/uploads/ebook/tdg/ADNVANCED%20MILITARY%20PLATFORM%20DESIGN/Principles%20of%20Naval%20Architecture%20(Second%20Revision),%20Volume%20I.pdf. Accessed 18 Apr 2025. This book provides a comprehensive overview of the principles in the field of naval architecture, including loads and structural analysis methods
7. K.-T. Ma, Y. Luo, C.T.T. Kwan et al., *Mooring System Engineering for Offshore Structures* (Elsevier, Amsterdam, 2019). https://www.sciencedirect.com/book/9780128185513/mooring-system-engineering-for-offshore-structures. Accessed 16 Apr 2025. This book provides a comprehensive look at the fundamentals involved during mooring designs with coverage on current standards and codes, mooring analysis, and more
8. C.V. Oldershausen, DNV RU-NAV provides common entry point for naval rules (2025). https://www.dnv.com/news/dnv-ru-nav-provides-common-entry-point-for-naval-rules-200589/. Accessed 15 Apr 2025. This article discusses the implementation of the ship type naval in Naticus Hull (NH20), a structural assessment tool
9. M. Pricop, V. Oncica, I.C. Scurtu, Factor of safety in offshore structures design according to environmental loads. "Mircea cel Batran" Nav. Acad. Sci. Bull. **XVI**(2), 152–157 (2013). https://www.researchgate.net/publication/301553578_FACTOR_OF_SAFETY_IN_OFFSHORE_STRUCTURES_DESIGN_ACCORDING_TO_ENVIRONMENTAL_LOADS. Accessed 15 Apr 2025. This paper discusses the design requirements for floating platforms to operate safely under environmental loads

Chapter 5
Naval Stability Calculation

Abstract The chapter "Naval Stability Calculation" explores the fundamental principles governing a ship's equilibrium, buoyancy, and resistance to capsizing under various conditions. It begins by defining equilibrium in ship stability, where gravitational and buoyant forces are balanced, and introduces Archimedes' Principle as the foundation for buoyancy calculations. The text details how the center of gravity (G), center of buoyancy (B), and metacenter (M) interact to determine whether a vessel is in stable, neutral, or unstable equilibrium, emphasizing the importance of metacentric height (GM) as a measure of stability. It also discusses hydrostatic pressure and its role in generating buoyant force, along with the influence of factors such as loading, damage, and free surface effect on stability. The chapter further examines transverse and longitudinal stability, providing mathematical formulations for metacentric radius, metacentric height, and center of flotation, and includes example calculations for practical understanding. Finally, it introduces the concepts of moment of inertia and the parallel axis theorem as tools for analyzing the ship's resistance to rolling and pitching, concluding with composite examples that illustrate how geometric properties influence a vessel's overall stability.

5.1 Equilibrium in Ship Stability

In naval engineering, equilibrium in ship stability is a key idea. When all forces and moments operating on the ship are balanced out, a stable condition results in which the ship keeps its position and orientation. This equilibrium can be impacted by a number of variables, including the load distribution on the ship and the environment.

M. Vinicius da Silva Neves et al., *Offshore Operations Engineering*,
SpringerBriefs in Applied Sciences and Technology,
https://doi.org/10.1007/978-3-032-04715-1_5

5.2 Buoyancy and Archimedes' Principle

Buoyancy is a fundamental concept in naval stability. It is the force that allows a ship to float in water. This force is governed by Archimedes' Principle, which states that any object, wholly or partially immersed in a fluid, is buoyed up by a force equal to the weight of the fluid displaced by the object. The formula for buoyancy (B) is:

$$B = \rho * g * V$$

where:

- B is the buoyant force,
- ρ is the density of the fluid,
- g is the acceleration due to gravity, and
- V is the volume of the displaced fluid.

5.2.1 Archimedes' Principle

The mathematician and scientist Archimedes of Syracuse created the Archimedes' Principle, which is a basic idea in naval construction and maritime engineering. According to this, a body submerged in a stagnant fluid experiences a force that is vertically upward and proportional to the weight of the displaced fluid.

This idea can be stated mathematically as:

$$F = \rho * g * V$$

where:

- F is the buoyant force,
- ρ is the fluid density,
- V is the volume of the displaced fluid, and
- g is the acceleration due to gravity.

5.2.2 Forces and Moments in Equilibrium

In the context of ship stability, there are several forces and moments that contribute to the equilibrium state:

1. **Gravitational Force**: This is the weight of the ship acting downwards through the center of gravity (G). The location of G is determined by the distribution of weight within the ship.

2. **Buoyant Force**: This is the force exerted by the water, acting upwards through the center of buoyancy (B). The location of B depends on the shape of the ship's hull and the waterline.
3. **Hydrostatic Righting Moment**: When a ship is inclined, the buoyant force and gravitational force act along different lines, creating a moment that tends to restore the ship to an upright position. This is known as the righting moment.

The equilibrium of a ship is achieved when the sum of all forces and moments acting on the ship is zero. This means that the gravitational force equals the buoyant force, and the sum of all moments about any point is zero.

5.2.3 Stability Conditions

The stability of a ship in equilibrium can be described in terms of the metacentric height (GM), which is the distance between G and the metacenter (M). The metacenter is the point where the line of action of the buoyant force shifts vertically concerning the point of flotation, when the ship is inclined at small angles.

1. **Stable Equilibrium**: If M is above G (GM is positive), the ship is in stable equilibrium. Any inclination will generate a righting moment that tends to restore the ship to an upright position.
2. **Neutral Equilibrium**: If M coincides with G (GM is zero), the ship is in neutral equilibrium. The ship will not tend to return to an upright position or to capsize further after an inclination.
3. **Unstable Equilibrium**: If M is below G (GM is negative), the ship is in unstable equilibrium. Any inclination will generate a moment that tends to increase the angle of heel, potentially leading to capsize.

5.2.4 Equilibrium in Various Conditions

Different circumstances, such as loading, damage, and environmental factors, can impact a ship's equilibrium. For instance, overloading or inappropriate loading can reduce the center of gravity, increasing stability but possibly allowing the ship to float dangerously close to the surface of the sea. The ship's center of gravity may increase due to damage that allows water to enter, which can decrease stability. Wind and waves, for example, might cause the ship to experience moments and forces that change its equilibrium.

5.2.5 *Equilibrium and Ship Design*

Understanding and maintaining equilibrium is crucial in ship design and operation. Naval architects must design ships with a suitable metacentric height to ensure stability under expected operating conditions. Ship operators must manage loading and ballasting to maintain stability in various conditions.

5.3 Hydrostatic Pressure

Hydrostatic pressure is the pressure exerted by a fluid at equilibrium due to the force of gravity, or in other words, it is the pressure difference between the bottom and the top of the hull that effectively results in the buoyant force. It increases linearly with increasing depth, and for a fluid element in equilibrium, the balance of vertical forces yields the following formula for hydrostatic pressure:

$$p = \rho * g * h$$

where:

- p is the hydrostatic pressure,
- ρ is the fluid density,
- g is the acceleration due to gravity, and
- h is the height of the fluid column above the point in question.

In the context of ship stability, hydrostatic pressure plays a crucial role. It is the key factor that determines the buoyant force acting on the ship, which in turn affects the ship's ability to maintain stability and resist capsizing.

5.3.1 *Example: Calculating Hydrostatic Pressure*

Let's consider a simple example. Suppose we have a ship that is floating in seawater. The density of seawater is approximately 1025 $\mathrm{kg/m^3}$, and the acceleration due to gravity is approximately 9.81 $\mathrm{m/s^2}$. If we want to calculate the hydrostatic pressure 10 m below the surface, we can use the formula for hydrostatic pressure:

$$p = \rho * g * h$$

$$p = 1025\,\mathrm{kg/m^3} * 9.81\,\mathrm{m/s^2} * 10\,\mathrm{m} \quad p = 100{,}512.5\,\mathrm{Pa}$$

So, the hydrostatic pressure 10 m below the surface is approximately 100,512.5 Pa.

5.4 Center of Area, Center of Mass, and Center of Gravity

The center of area, center of mass, and center of gravity are all crucial points in a ship's structure:

The **Center of Area** (also known as the centroid) is the point in a plane figure that is equidistant from all its sides. It's the point at which the figure would balance if it were cut out of a sheet of material. For a simple geometric shape, the centroid can be found using basic geometric formulas. For example, the centroid of a rectangle is simply at the intersection of its diagonals. For more complex shapes, the centroid can be found by dividing the shape into simpler parts, finding the centroid of each part, and then finding the weighted average of these points, where the weights are the areas of the parts.

The **Center of Mass** is the unique point in an object or system of particles where the weighted relative position of the distributed mass sums to zero. It's the balance point of the object. The center of mass of a system of particles is the weighted average of their positions, where the weights are the masses of the particles. In formula terms, if you have n particles with positions $r1$, $r2$, ..., rn and masses $m1$, $m2$, ..., mn, the center of mass R is given by:

$$R = \frac{(m1r1 + m2r2 + \cdots + mn * rn)}{(m1 + m2 + \cdots + mn)}$$

The **Center of Gravity** is the point in an object around which the resultant torque due to gravity forces vanishes. It's the point at which the weight of the body acts vertically downward. The center of gravity is essentially the same as the center of mass when the gravitational field is uniform. In a non-uniform field, the center of gravity can be found by considering the weight of each part of the object instead of its mass in the formula for the center of mass.

5.5 Stability: Transverse and Longitudinal

Stability in naval architecture refers to the ability of a ship to return to its original position after being disturbed. There are two types of stability: transverse and longitudinal.

Longitudinal Stability (also known as pitch stability) refers to the stability of a ship in the fore-and-aft (front-to-back) direction. It's determined by the distribution of weight along the length of the ship.

Transverse Stability refers to the stability of a ship in the athwartships (side-to-side) direction. The key to transverse stability is the relationship between the center of gravity (G) and the center of buoyancy (B). When a ship heels (tilts to one side), the center of buoyancy shifts to the side. If B is above G, the ship is stable and will return to an upright position. If B is below G, the ship is unstable and will capsize.

Transverse stability refers to the stability of a ship in the athwartships (side-to-side) direction. It is determined by the relationship between the center of gravity (G) and the metacenter (M).

5.5.1 Stable, Neutral, and Unstable Conditions

- **Stable Condition**: If M is above G (GM is positive), the ship is in a stable condition. Any inclination will generate a righting moment that tends to restore the ship to an upright position.
- **Neutral Condition**: If M coincides with G (GM is zero), the ship is in a neutral condition. The ship will not tend to return to an upright position or to capsize further after an inclination.
- **Unstable Condition**: If M is below G (GM is negative), the ship is in an unstable condition. Any inclination will generate a moment that tends to increase the angle of heel, potentially leading to capsize.

5.5.2 Metacentric Radius and Metacentric Height

The metacentric radius (BM) is a measure of the ship's shape and is given by the formula:

$$BM = \frac{I}{V}$$

where:

- I is the moment of inertia of the waterplane area about the longitudinal axis, and
- V is the volume of displacement.

The metacentric height (GM) is the distance between G and M and is given by the formula:

$$GM = KB + BM - KG$$

where:

- KB is the distance from the keel to the center of buoyancy,
- BM is the metacentric radius, and
- KG is the distance from the keel to the center of gravity.

5.5.3 *Calculation of* KM *for a Box Ship*

For a box-shaped ship, the metacentric height can be calculated using the formula:

$$KM = \frac{h}{2}$$

where:

- h is the draft of the ship.

5.5.4 *Deduction of the Metacentric Radius*

The metacentric radius can be deduced from the metacentric height and the positions of the keel, the center of buoyancy, and the center of gravity using the formula for GM given above.

5.5.5 *Center of Flotation*

The center of flotation is the point in the waterplane area through which the buoyant force acts. It is the centroid of the waterplane area. It is important to note that the center of flotation can shift depending on the ship's inclination state.

5.5.6 *Factors Affecting* KM

The value of KM is affected by the ship's breadth (B) and draft (T). In general, increasing the breadth or decreasing the draft will increase KM, making the ship more stable.

5.5.7 *Free Surface Effect*

The free surface effect refers to the decrease in stability caused by the movement of liquid in partially filled tanks or compartments. This movement shifts the center of gravity, reducing the metacentric height and hence the stability.

Please note that these are general principles and formulas. For a more accurate analysis of a specific ship, more complex methods and software tools are typically used.

5.5.8 Calculating Stability

The stability of a ship can be calculated using the metacentric height (GM), which is the distance between G and a point called the metacenter (M). The metacenter is the point where the line of action of the buoyancy force intersects the midline of the ship when it is tilted. The formula for GM is:

$$GM = KB + BM - KG$$

where:

- KB is the distance from the keel to the center of buoyancy,
- BM is the metacentric radius (an indicator of the ship's shape),
- KG is the distance from the keel to the center of gravity, and
- If GM is positive, the ship is stable. If GM is negative, the ship is unstable.

5.5.9 Example Calculation

Let's consider a simple example. Suppose we have a ship with the following characteristics:

- $KB = 5$ m
- $BM = 1$ m
- $KG = 4$ m.

We can calculate GM as follows:

$$GM = KB + BM - KG$$

$$GM = 5\,\text{m} + 1\,\text{m} - 4\,\text{m} = 2\,\text{m}$$

Since GM is positive, the ship is stable.

5.6 Inertia and Plane Figures in Ship Stability

Inertia is a fundamental concept in physics and engineering that describes an object's resistance to changes in its state of motion. In the context of ship stability, the concept of inertia is often applied to plane figures, such as cross-sections of the ship's hull, to calculate properties like the moment of inertia and the radius of gyration, which are important for understanding the ship's resistance to rolling and pitching.

5.6.1 Moment of Inertia

The moment of inertia of a plane figure about an axis is a measure of the figure's resistance to rotational motion about that axis. It is calculated by integrating the square of the distance from each infinitesimal element of the figure to the axis, weighted by the element's area. For example, the moment of inertia I of a rectangle of base b and height h about an axis through its centroid and perpendicular to the plane of the figure is given by:

$$I = \frac{(b * h^3)}{12}$$

5.6.2 Parallel Axis Theorem

The parallel axis theorem is a useful tool for calculating the moment of inertia of a plane figure about any axis, given the moment of inertia about a parallel axis through the figure's centroid. The theorem states that the moment of inertia about the parallel axis is equal to the moment of inertia about the centroidal axis plus the product of the figure's total area and the square of the distance between the two axes. If Ic is the moment of inertia about the centroidal axis, A is the figure's area, and d is the distance between the axes, then the moment of inertia I about the parallel axis is given by:

$$I = \text{Ic} + A * d^2$$

5.6.3 Example: Rectangle

Consider a rectangle with base $b = 2$ m and height $h = 1$ m. The moment of inertia about an axis through the rectangle's centroid and perpendicular to the plane of the rectangle is:

$$\text{Ic} = \frac{(b * h^3)}{12}$$

$$\text{Ic} = \frac{(2 * 1^3)}{12} = \frac{1}{6\,\text{m}^4}$$

Now suppose we want to find the moment of inertia about an axis parallel to the centroidal axis but 3 m away. Using the parallel axis theorem:

$$I = \text{Ic} + A * d^2$$

$$I = \frac{1}{6} + (2 * 1) * 3^2 = \frac{1}{6} + 18 = \frac{109}{6\,\text{m}^4}$$

This example illustrates how the parallel axis theorem can be used to calculate the moment of inertia of a plane figure about any axis, which is a crucial calculation in understanding a ship's resistance to rolling and pitching.

5.6.4 Example: Composite Plane

Let's consider a composite figure made up of a square, a triangle, and a rectangle. We'll calculate the moment of inertia of this composite figure about an axis perpendicular to the plane of the figure and passing through a certain point.

This example illustrates how the parallel axis theorem can be used to calculate the moment of inertia of a composite figure made up of simple geometric shapes.

Let's say the square has side length a, the triangle has base b and height h, and the rectangle has width w and height h. Let's also say that the axis passes through the centroid of the square and is d units away from the centroids of the triangle and the rectangle.

Square: The moment of inertia of the square about the axis is:

$$I_{\text{square}} = \frac{a^4}{12}$$

Since the axis passes through the centroid of the square, we don't need to use the parallel axis theorem for the square.

Triangle: The moment of inertia of the triangle about an axis through its centroid is:

$$I_{\text{triangle centroid}} = \frac{\left(b * h^3\right)}{36}$$

Using the parallel axis theorem, the moment of inertia of the triangle about the given axis is:

$$I_{\text{triangle}} = I_{\text{triangle centroid}} + A_{\text{triangle}} * d^2$$

$$I_{\text{triangle}} = \frac{\left(b * h^3\right)}{36} + \left(\frac{b * h}{2}\right) * d^2$$

Rectangle: The moment of inertia of the rectangle about an axis through its centroid is:

$$I_{\text{rectangle centroid}} = \frac{\left(w * h^3\right)}{12}$$

Using the parallel axis theorem, the moment of inertia of the rectangle about the given axis is:

$$I_{\text{rectangle}} = \frac{\left(w * h^3\right)}{12} + (w * h) * d^2$$

The moment of inertia of the composite figure about the axis is the sum of the moments of inertia of the individual figures:

$$I_{\text{composite}} = I_{\text{square}} + I_{\text{triangle}} + I_{\text{rectangle}}$$

Bibliography

1. B. Barrass, C.D.R. Derrett, *Ship Stability for Masters and Mates* (Butterworth-Heinemann, Oxford, 2012). This book explores all aspects of ship stability and ship strength, squat, and interaction and trim, as well as materials stresses and forces
2. R.S. Crenshaw, *Naval Shiphandling* (Naval Institute Press, Annapolis, 1974). This book, while not specifically about stability, provides a comprehensive understanding of handling naval ships, which requires understanding of stability principles
3. C.S. Moore, J.R. Paulling, *The Principles of Naval Architecture Series: Intact Stability* (The Society of Naval Architects and Marine Engineers, New Jersey, 2010). This book provides a comprehensive understanding of a ship's stability and strength
4. G.C. Nibous, Notes on hydrostatic pressure. J. Ocean Eng. Mar. Energy **2**, 105–109 (2016). https://doi.org/10.1007/s40722-015-0035-1
5. E.C. Tupper, *Introduction to Naval Architecture* (Butterworth-Heinemann, Oxford, 2013). This book provides a broad understanding of naval architecture, including ship stability

Chapter 6
Meteorological and Oceanographic Considerations for Offshore Oil Production Operations

Abstract The chapter "Meteorological and Oceanographic Considerations for Offshore Oil Production Operations" examines how atmospheric and oceanic phenomena affect the safety, efficiency, and sustainability of offshore oil activities. It discusses the main environmental factors—waves, tides, currents, and winds—that influence platform design, positioning, and operation. The chapter distinguishes between wind waves and swell, emphasizing their effects on structural loads, freeboard, risers, and deck drainage systems. It explains the formation and impact of tides, particularly their role in platform mooring and foundation design, as well as the significance of ocean currents driven by wind, temperature, and salinity gradients, which affect both structural stresses and spill dispersion. The section on winds details their generation through pressure differences and the Coriolis effect, highlighting their relevance for structural stability, operational safety, and compliance with IMO standards. Further, it explores cold fronts, troughs, and pressure systems in the Southern Hemisphere, explaining how their interactions generate storms and shifts in weather patterns that directly impact offshore operations. The chapter concludes by analyzing the Intertropical Convergence Zone (ITCZ) and South Atlantic High-Pressure System, along with cloud formations and storm dynamics, stressing that a thorough understanding of meteorological and oceanographic conditions is vital for safe, efficient, and resilient offshore oil production.

The safety, effectiveness, and sustainability of offshore oil production operations depend on an understanding of the meteorological and oceanographic phenomena that influence these operations. This chapter will focus on the key elements of meteorology and oceanography that impact offshore oil production, namely waves, tides, currents, and winds.

6.1 Wind Waves and Swell

One of the most important elements affecting offshore constructions is wave action.

M. Vinicius da Silva Neves et al., *Offshore Operations Engineering*,
SpringerBriefs in Applied Sciences and Technology,
https://doi.org/10.1007/978-3-032-04715-1_6

It is common in the maritime culture to refer to sea waves by dividing them into two types: wind waves and swell waves.

"Wind waves" are short and irregular ocean waves, primarily generated by local winds. The frictional drag of the wind over the sea's surface causes waves to form. When the local wind blows across the surface of the water, it transfers energy to the water and creates a chaotic surface, resulting in short and intermittent periods between the waves. It's common to hear the expression "ripples" as well, which refers to a type of wind wave that is smaller, smoother, and generally occurs in shallower areas or in response to gentle winds, representing a smaller and less intense scale.

In contrast, "swell waves" refer to a series of long-wavelength ocean waves that have traveled away from their area of generation, typically caused by sustained winds over a large area for a long duration, including the influence of storms in the open ocean. Unlike local wind waves, swells are characterized by their smoother, more regular shape and longer periods between waves. In comparison to wind waves, swell waves can propagate over great distances across the ocean.

Factors related to the height, duration, and direction of both types of waves can have a significant impact on how offshore oil rigs are built and function.

When designing offshore constructions, wave period and wave height (see Fig. 6.1) are both an important consideration. Waves periods are commonly used to design a maritime structure, as they define the distance between their troughs and crests, which can pass by the maritime structures, causing significant stresses. Wave height establishes the necessary freeboard and significantly influences the designs of riser and mooring systems for the platform, and additionally is essential to incorporate features such as scuppers designed to drain green water (wave-induced water onboard the vessel) from the decks; projects for hatches that ensure the watertightness of specific decks during water onboard situations; the calculation of the vessel's reserve buoyancy that is necessary to support the weight of wave-induced water onboard; and load equipment design for prevent cargo from being washed overboard during adverse sea conditions.

Considering all these factors is essential, as they can impact on the platform's dynamic response and the effectiveness of oil production activities, as well as the overall safety and operational integrity of offshore structures.

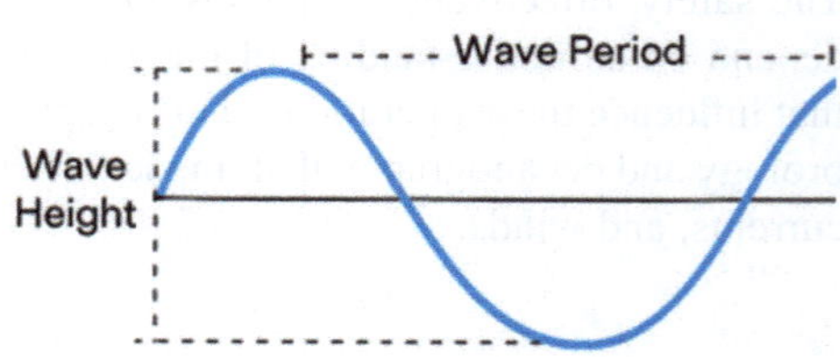

Fig. 6.1 Wave height and period

6.2 Tides

The gravitational pull of the moon and sun generate tides, which are the regular rise and fall of sea levels. Tidal changes can have a big impact on offshore operations, especially in areas with shallow water.

The gravitational interplay of the Earth, moon, and sun results in tides. The ocean on the other side of the Earth, in the moon's direction, bulges out due to the moon's gravitational attraction. High tides caused these bulges. Low tide occurs in the regions between high tides. Semidiurnal tides, which have two high tides and two low tides each day, and diurnal tides, which have one high tide and one low tide each day, are the two basic types of tides. Although the sun has an impact on tides as well, it has a smaller impact than the moon due to the sun's greater distance from the earth.

The substructure position and the foundation of the platform's structure, in particular, must take tidal ranges into account. The positioning and mooring of the platform may also have some impact from the tidal currents, which are horizontal water motions brought on by tides.

6.3 Currents

Ocean currents, which are the ongoing movements of seawater caused by variations in wind, temperature, and salinity, can significantly affect offshore oil extraction activities.

Wind, temperature variations (thermohaline circulation—THC), and the Coriolis effect resulting from the Earth's rotation are some of the elements that influence ocean currents. Surface currents and deep ocean currents are the two main categories of currents. About 10% of ocean water is made up of surface currents, which include the Gulf Stream in the Atlantic and the Kuroshio Current in the Pacific. Surface currents are predominantly influenced by wind. The remaining 90% of ocean water is made up of deep ocean currents, commonly referred to as the "global conveyor belt," which are propelled by variations in water density.

On offshore structures, currents can impose significant stresses and impact the movement and dispersion of released or spilled oil. They also have a significant influence on diving operations, which may need to be considered, as some of these operations are predicted to occur to meet regulatory and certification requirements. Designing the platform and organizing emergency response plans require a thorough understanding of the patterns and behavior of ocean currents.

6.4 Winds

Offshore oil platforms are designed and operated with wind as a primary consideration. Given its significant importance and potential influence on maritime structures, the consideration of wind in designs is a requirement outlined in the rules of classification societies and the guidelines of the International Maritime Organization (IMO), such as the SOLAS Convention.

In addition to the stability of maritime structures, the safety of activities, particularly cargo handling operations, can be impacted by wind forces.

Differences in air pressure are what generate winds. When there are high and low pressure zones, air moves from the high-pressure zone to the low-pressure zone, resulting in wind. The Coriolis effect is responsible for deflecting the movement of the wind, causing it to rotate around high and low-pressure regions. In the Northern Hemisphere, this movement is deflected to the right, resulting in a counterclockwise rotation around low-pressure areas and a clockwise rotation around high-pressure areas, while in the Southern Hemisphere, the deflection occurs to the left, leading to a clockwise rotation around low-pressure areas and a counterclockwise rotation around high-pressure areas.

Winds can also be caused by fronts, which are the boundaries between air masses with varying temperatures. When a cold air mass pushes under a warm air mass, the warm air rises and may eventually produce clouds and precipitation. This is how cold fronts form.

Additionally, wind speed and direction can affect wave conditions and, as a result, the dynamic response of the platform. For operational planning, as well as for the design of the platform and its components, accurate wind forecasting is crucial.

6.5 Cold Fronts and Troughs in the Southern Hemisphere

The leading edge of a cooler mass of air that is replacing a warmer one is a cold front. A frontal transition is the term for this action. Cold fronts normally flow from the southwest to the northeast in the Southern Hemisphere, but the larger weather pattern can affect their orientation.

When a cold air mass pushes into a warmer air mass, a cold front is formed. Due to the fact that cold air is denser than warm air, it slips on top of the warm air. As it is lifted, the warm air gets cooler. If there is enough moisture in the elevated air, clouds will form and precipitation may happen.

Cold fronts are shown on weather maps as a solid line with triangles pointing in their direction of motion. A cold front leaves the air behind it substantially colder and drier than the air in front of it.

Interactions Between High- and Low-Pressure Regions

The movement of air in the atmosphere and the formation of weather patterns are both governed by the interaction of high- and low-pressure zones. High-pressure regions give way to low-pressure regions when air moves. The air flows more quickly when there is a bigger pressure difference.

In the Southern Hemisphere, air travels anticlockwise around low-pressure systems and clockwise around high-pressure systems. This results from the Coriolis effect, which is brought on by the rotation of the Earth.

The air from the high-pressure system will rush toward the low-pressure area as it comes into contact with a low-pressure system, creating winds. The high-pressure system will "fill" a low-pressure system that is a deep depression, which will eventually cause the low-pressure system to dissipate.

Weather fronts, which are the boundaries between air masses of various temperatures and humidities, can emerge as a result of the passage of air from high to low pressure. The interaction of the air masses along these fronts can result in precipitation and storms.

Troughs in Meteorological Charts

A trough, or a "cavado" in Portuguese, is an elongated area of relatively low atmospheric pressure. It is associated with and extends from a low-pressure region. On a weather chart, troughs can be identified as a line, or an elongated area, where the isobars are kinked.

Troughs can bring clouds, showers, and a change in wind direction. The weather usually clears after the trough passes. In the Southern Hemisphere, winds ahead of the trough typically come from the north (warmer, bringing rain), and winds behind the trough usually come from the south (cooler, bringing clearer weather).

6.6 The Intertropical Convergence Zone and High-Pressure Systems

Located close to the equator, the Intertropical Convergence Zone (ITCZ), also called the "Zona de Convergência Intertropical" in Brazil, is a region of low pressure. It is the location of the "convergence" of the trade winds between the Northern and Southern Hemispheres. Due to the air being forced upward by this convergence, clouds and frequent thunderstorms are created.

Depending on how the Earth is tilted in reference to the sun, the ITCZ's position can shift seasonally, migrating either north or south of the equator. The South Atlantic High-Pressure System, which is a key component of the atmospheric circulation over the South Atlantic Ocean, is one example of a high-pressure system whose location and intensity are influenced by this movement.

Off the eastern coast of Brazil, in the subtropics, is where the South Atlantic High is often found. The South Atlantic High can stretch further north during the summer in the Northern Hemisphere when the ITCZ is situated north of the equator, affecting the weather along the Brazilian coast. On the other hand, the South Atlantic High likewise moves southward when the ITCZ changes south of the equator during the summer in the Southern Hemisphere.

The South Atlantic High interacts with the ITCZ and has a substantial impact on wind patterns, sea surface temperatures, and rainfall patterns in the area, all of which can have a big impact on offshore activities.

6.7 Storms and Cloud Types

Temperature variations, atmospheric instability, and weather fronts are just a few of the variables that can lead to storms in the ocean. They may cause strong gusts, copious amounts of precipitation, and dangerous sea conditions.

Clouds are an important component of the weather and can reveal many weather patterns. Cumulus clouds are puffy clouds that occasionally resemble floating pieces of cotton and are frequently connected to pleasant weather. Tall thunderstorm clouds known as cumulonimbus can extend into the troposphere. The sky is covered in a covering of stratus clouds, which frequently causes cloudy skies and the possibility of light precipitation. Cirrus clouds are ice crystal-filled, thin, wispy clouds that signal fair to pleasant weather.

As a result, for offshore oil production facilities to operate safely and effectively, a thorough understanding of meteorological and oceanographic phenomena is required. For offshore oil production to be sustainable and profitable, these issues should be carefully taken into account during platform design and operational planning.

Note: This chapter is based on the principles outlined in the book "Meteorologia e Oceanografia: Usuário Navegante" by Paulo Roberto Valgas Lobo [1]. For a more detailed understanding of these topics, it is recommended to refer to the original text.

Reference

1. P.R.V. Lobo, *Meteorologia e Oceanografia: Usuário Navegante*, 4th edn. (Vozes, Petrópolis, 2019)

Chapter 7
Basic Notions of Navigation and Maneuvering at Sea

Abstract The chapter "Basic Notions of Navigation and Maneuvering at Sea" provides a comprehensive overview of the essential principles of maritime navigation and their critical role in offshore oil production operations. It begins by classifying navigation into oceanic, coastal, and restricted waters, outlining their specific precision requirements and position-fixing frequencies. The text then explores the types of navigation techniques—celestial, visual, electronic, and dead reckoning—highlighting how they ensure safe vessel movement in various marine environments. A detailed explanation of the Earth's geometry, including great and small circles, geographic coordinates, and map projections (especially the Mercator projection), establishes the foundation for chart-based navigation. The chapter also defines bearings (true, magnetic, and compass) and explains how magnetic declination and deviation affect course plotting. It introduces the units of measurement used in navigation, such as nautical miles, knots, and angular degrees, and describes a wide range of navigational instruments, from traditional compasses and sextants to modern GPS and ECDIS systems. Additionally, it summarizes the International Regulations for Preventing Collisions at Sea (COLREGs) and their role in ensuring maritime safety through standardized vessel behavior, lights, and signals. Finally, the chapter connects navigation fundamentals to offshore oil operations, emphasizing their application in oil offloading, water and diesel transfers, cargo movements, diving activities, and simultaneous offshore operations. It stresses the importance of dynamic positioning, drift control, environmental monitoring, communication, and regulatory compliance to ensure safety, efficiency, and environmental protection in complex offshore environments.

Navigation is the practice of properly directing a ship from one location to another on the Earth's surface and is essential in offshore engineering activities for oil extraction. This chapter explores the fundamental elements of navigation, including various types of navigation, the conceptual framework of the Earth's sphere, the geographic coordinate system, units of measurement in navigation, navigation instruments, and the International Regulations for Preventing Collisions at Sea (COLREGs).

M. Vinicius da Silva Neves et al., *Offshore Operations Engineering*,
SpringerBriefs in Applied Sciences and Technology,
https://doi.org/10.1007/978-3-032-04715-1_7

7.1 Types of Navigation

Navigation is typically classified into three primary groups based on the proximity to the nearest coast or danger:

- **Oceanic Navigation**: Conducted in open sea, typically beyond 50 nautical miles from the coast and in water depths exceeding 200 m. Precision of 1–2 miles is acceptable due to the distant hazards. The frequency of position fixing to be plotted on a nautical chart should be at least three times a day during oceanic navigation.
- **Coastal Navigation**: Occurs closer to shore, within 3–50 nautical miles of the coast, where water depths range from 20 to 200 m. A higher precision of 0.1 mile (or 200 yards) is required due to the proximity of hazards. Position fixing frequency for coastal navigation should be at least every 30 min.
- **Restricted Waters Navigation**: The most cautious navigation, taking place in ports, harbors, bays, canals, lakes, and rivers, where navigation is constrained by various factors. This type requires precise navigation (to the order of 0.05 miles or 100 yards) due to the very close proximity of hazards. The frequency of position fixing in restricted waters should be at least every 3 min.

Mariners can use many strategies to establish their position at sea when navigating these environments:

- **Celestial Navigation**: Utilizes observations of celestial bodies (the sun, moon, planets, and stars) to calculate one's position on the globe. This ancient and reliable method is particularly useful for oceanic navigation where no landmarks are visible.
- **Visual Navigation**: Involves using visible landmarks, aids to navigation (such as lighthouses and buoys), and depth soundings. This method is most applicable to coastal and restricted waters navigation, where such references are available.
- **Electronic Navigation**: Employs a range of technologies including GPS (Global Positioning System), radar, and electronic charts (ECDIS—Electronic Chart Display and Information System) to determine the vessel's position with high accuracy. Electronic navigation is essential across all types of navigation but is especially critical in ensuring safety in restricted waters.
- **Dead Reckoning**: Estimates the current position based on previously determined positions, and known or estimated speeds over elapsed time, and course. Dead reckoning is used in conjunction with other navigation methods to provide a continuous position estimate between fixes.

Proper utilization of these navigation techniques ensures the secure and effective movement of ships in various marine environments.

7.2 Navigation Bearings and Earth's Sphere: Lines, Points, and Planes

To accurately determine a vessel´s position on the ocean, it is crucial to understand how the Earth is represented on maps. Additionally, knowledge of bearing concepts is essential to plot a navigation course and follow the correct path to reach a destination. Several lines can be drawn on the Earth's surface to assist in measuring distances and defining routes. Notable examples include:

- **Great Circles**: The largest circles that can be drawn on the globe, such as the Equator and Meridians, which are essential for defining geographic coordinates. Great circles result from the intersection of the Earth's surface with a plane that passes through the Earth's center. They are used in navigation to define the shortest path between two points on the surface of the Earth.
- **Small Circles**: Circles that are smaller than the Earth's circumference, such as parallels of latitude. Small circles result from the intersection of the Earth's surface with a plane that does not pass through the Earth's center.
- **Equator**: The primary great circle that divides the Earth into the Northern and Southern Hemispheres. It is the baseline from which latitude is measured.
- **Meridians**: Great circles passing through the poles, used to define longitude. Each meridian is half of a great circle and intersects the equator at right angles. Longitude is conventionally measured from the Prime Meridian, which passes through Greenwich, England.

Maps and charts are representations of the Earth's surface that serve different purposes:

- **Maps**: Represent the Earth or sections of its surface on a flat plane, showing political boundaries, physical features, cities, and other geographical, social, or economic information. Maps are not suitable for navigation due to the distortions inherent in projecting a spherical surface onto a plane.
- **Charts**: Specifically designed for navigation, nautical charts represent parts of the Earth's surface with a focus on navigable waters. They include information on coastlines, ports, channels, sea hazards, currents, depths, navigation aids, and more. Charts are crafted to facilitate the resolution of navigational problems, where the main elements are angles and distances, or the determination of position using geographic coordinates (latitude and longitude).

7.2.1 Desirable Conditions for Earth's Surface Representation

Ideally, a representation of the Earth on a flat surface would maintain the true shapes of areas (islands, continents, etc.), the unaltered dimensions of these areas, and allow for the use of straight lines to represent navigation courses. Additionally, it would enable easy determination of geographic coordinates without extensive calculations. However, achieving all these conditions is impossible in a flat projection of our planet.

Given the impracticality of using a globe for navigation, nautical charts are employed instead. It is desirable for a nautical chart to satisfy one or two of the following properties:

- **Equidistant**: The chart does not distort distances.
- **Equivalent**: The chart does not distort areas.
- **Conformal**: The chart maintains angles, meaning that shapes are preserved over small areas.

The most commonly used projection in maritime navigation is the Mercator projection. Nautical charts created using the Mercator projection employ rhumb lines (loxodromes), representing courses as straight lines at constant angles. This cylindrical equatorial conformal projection was devised by Gerhard Kremer (known as Mercator) in 1512. It represents the Earth on a plane in such a way that courses plotted as straight lines correspond to constant compass bearings, facilitating navigation.

Advantages of Mercator charts include:

- Meridians are represented as vertical straight lines, and parallels and the equator as horizontal straight lines, perpendicular to the meridians.
- Cardinal points (N–S–E–W) are easily identified on a Mercator chart.
- Geographic coordinates of any point can be easily determined (latitudes in degrees and minutes are represented vertically, and longitudes in degrees and minutes horizontally).
- Angles measured in navigation are represented by identical angles on the chart.
- Distances and directions can be measured directly on the charts.

The "International Chart of Standard Symbols and Abbreviations" (known as Chart 12.000, published by DHN) facilitates the use of nautical charts by standardizing symbols and abbreviations. However, care must be taken when reading instructions contained in foreign nautical charts to avoid errors.

7.2.2 Geographic Coordinate System

The geographic coordinate system allows for the precise determination of positions on the Earth's surface using latitude and longitude:

- **Latitude**: Defined as the arc of the meridian between the Equator and the parallel of a specific location, measured from 0 to 90° north or south of the Equator. Latitude is a key component in determining a vessel's position on the Earth's surface.
- **Average Latitude**: Calculated as the mean parallel between the parallels passing through two points. Depending on whether the points are in the same hemisphere

or not, the average latitude is obtained by the semi-sum or semi-difference of the latitudes of the two locations.

- **Longitude**: Defined as the arc of the Equator between the Prime Meridian (Greenwich Meridian) and the meridian of a specific location, measured from 0 to 180° east or west of the Prime Meridian. Longitude, together with latitude, enables precise location identification on the globe.

7.2.3 Navigation Bearings

In the realm of maritime navigation, understanding the various types of bearings is crucial for accurately determining a vessel's course and position. Bearings are defined as the direction from one point to another, typically measured in degrees from a reference direction. The primary types of bearings used in navigation are:

- **True Bearing**[1] **(TB)**: The angle measured from True North (TN) to the direction of interest, such as the bow of the ship or a distant object. True North is the direction along the Earth's surface toward the geographic North Pole.
- **Magnetic Bearing (MB)**: The angle measured from Magnetic North (MN) to the direction of interest. Magnetic North is the direction toward the Earth's magnetic north pole, which varies based on location and changes over time due to geomagnetic changes.
- **Compass Bearing (CB)**: The angle measured from Compass North (CN) to the direction of interest. Compass North is determined by the direction in which a magnetic compass needle points, which can vary from Magnetic North due to local magnetic anomalies and the specific characteristics of the compass.

The angular discrepancy between True North and Magnetic North, known as magnetic declination, is caused by the Earth's magnetic field not aligning completely with its geographic poles. The declination might be either eastward or westward, based on the relative positions of Magnetic North and True North at the individual location. Magnetic declination is essential for converting real bearings to magnetic bearings.

Additionally, the compass bearing can vary from the magnetic bearing because of magnetic deviation, caused by magnetic fields produced by the ship or neighboring magnetic materials. Each compass on a ship can exhibit individual deviation caused by its specific placement and the magnetic conditions in its vicinity.

Mariners must be proficient in converting between several types of bearings to navigate effectively. The procedure includes compensating for magnetic declination and compass variation to guarantee that the course plotted on a map directs toward the desired destination. The fundamental guidelines for converting between bearings are:

[1] Bearings are conventionally measured clockwise from True North, to ensure precision for all readers.

- **Magnetic Declination (Decl)** is always counted from True North toward the east (E) or west (W).
- **Deviation (*D*)** is always counted from Magnetic North toward the east (E) or west (W).
- **Magnetic Declination** can be directly obtained from the compass rose on the nautical chart where navigation is taking place. Given the chart's publication date, the indicated declination may need to be adjusted for the current year.

7.2.4 Tips for Plotting Bearings

When plotting the ship's position during a navigation problem, the drawn bow of the ship serves as the reference point for plotting True North, Magnetic North, and Compass North. This fixed point allows for the plotting of the origins of various bearings (True, Magnetic, and/or Compass).

Navigation problems are based on knowing or being given three out of five elements (True Bearing, Magnetic Bearing, Compass Bearing, the area's Magnetic Declination, and Deviation) and calculating the missing two.

In summary, when working on the chart, it's necessary to convert the true bearing to the compass bearing to guide the helmsman. Conversely, when working with the compass, it's essential to convert compass bearings back to true bearings for plotting on the nautical chart.

Understanding and correctly applying these concepts ensures safe and accurate navigation, allowing mariners to chart their course through the complexities of the Earth's magnetic environment.

7.3 Units of Measurement in Navigation

Navigation primarily uses the nautical mile, defined as one minute of latitude. Other units include knots (nautical miles per hour) for speed, and degrees and minutes for angular measurements.

Navigation relies on specific units of measurement that are essential for accurately determining position, distance, and speed at sea. The primary units include:

- **Nautical Mile**: The fundamental unit of distance in navigation, defined as one minute of latitude. This measurement is based on the Earth's circumference and is internationally recognized as equal to 1852 m or approximately 2000 yards. The nautical mile is used because it represents one minute of a degree of latitude, making calculations related to the Earth's surface more direct and straightforward for navigators.
- **Knot**: The unit of speed in navigation, defined as one nautical mile per hour. The term "knot" originates from the historical practice of measuring speed at sea using a chip log, which consisted of a wooden panel attached to a line with knots tied

at uniform intervals. As the ship moved, the line would play out, and the speed would be calculated based on the number of knots that passed in a fixed period. Today, the knot remains the standard measurement of speed on water, ensuring consistency in navigation and communication.

- **Degree and Minute**: Angular measurements are crucial for determining latitude and longitude, which are the coordinates used to specify locations on the Earth's surface. A degree is divided into 60 min (denoted as $60'$), and each minute can be further divided into 60 s (denoted as $60''$). These measurements allow for precise determination of position in terms of latitude (north or south of the equator) and longitude (east or west of the Prime Meridian at Greenwich).

For practical purposes, navigators use these units to calculate distance traveled (D) as a function of speed (V) and time (T), using the formula $D = V \times T$. This formula is fundamental in planning and executing navigation routes, ensuring that vessels reach their destinations efficiently and safely.

Understanding and accurately applying these units of measurement are crucial for all aspects of maritime navigation, from plotting courses and determining positions to estimating arrival times and fuel consumption.

7.4 Navigation Instruments

Navigators employ a diverse array of instruments to ascertain their position, direction, and speed, ensuring safe navigation across the oceans. These instruments are categorized based on their primary functions.

7.4.1 Direction Indicators

- **Magnetic Compasses**: Traditional navigation tools that align with the Earth's magnetic field, indicating magnetic north. Standard Magnetic Compasses are installed away from electromagnetic fields to prevent interference.
- **Gyroscopic Compasses**: Utilize the Earth's rotation to maintain alignment with True North, unaffected by magnetic anomalies. Gyrocompasses and their repeaters are found in various ship compartments, providing accurate heading information.

7.4.2 Speed and Distance Indicators

- **Log Devices (Odometers)**: Measure the distance traveled by a ship over water. Types include:

- **Surface Logs**: Measure speed at the water's surface but can be affected by sea conditions and debris.
- **Bottom Logs**: Measure speed relative to the sea floor, using pressure differences or electromagnetic principles.
- **Doppler Logs**: Utilize the Doppler Effect to measure speed over the ground, offering accuracy even at low speeds and unaffected by currents or tides.

7.4.3 Distance Indicators at Sea

- **Radar**: Employs electromagnetic waves to detect objects and measure distances, crucial for collision avoidance and navigation.
- **Laser Range Finders**: Use laser technology to measure precise distances to objects, though less common in maritime navigation.

7.4.4 Depth Indicators

- **Echo Sounders**: Utilize sound waves to measure the depth of water beneath a vessel, essential for avoiding underwater hazards and navigating shallow waters.

7.4.5 Vision and Drawing Enhancement Instruments

- **Telescopic Alidades**: Used for taking precise bearings of distant objects using a sight or telescope.
- **Azimuth Mirrors**: Assist in taking accurate bearings of celestial bodies for navigation.
- **Handheld Digital Magnetic Compasses**: Offer portable solutions for taking magnetic bearings.

7.4.6 Miscellaneous Instruments

- **Binoculars and Telescopes**: Enhance visual range for navigators.
- **Parallel Rulers and Dividers**: Aid in chart plotting and course planning.
- **Sextants**: Measure the angle between celestial objects and the horizon for celestial navigation.

7.4.7 Electronic Navigation Aids

- **GPS (Global Positioning System)**: Provides precise location data using satellites, crucial for modern navigation.
- **AIS (Automatic Identification System)**: Enhances vessel visibility and safety by automatically exchanging ship data with other nearby ships and coastal authorities.
- **ECDIS (Electronic Chart Display and Information System)**: Digital charting system that integrates GPS and other navigational sensors for real-time positioning and navigation planning.

7.4.8 Wind Measurement

- **Anemometers**: Measure wind speed and direction. It's important to note that anemometers measure apparent wind (wind experienced relative to the moving ship) rather than true wind.

Technological advancements have greatly improved the safety and efficiency of maritime navigation. Navigational instruments have advanced from the old magnetic compass to modern GPS and other systems, offering mariners precise and reliable navigation aids for different conditions. When combined, these tools enable effective voyage planning and execution, assuring the safety of the ship, crew, and cargo.

7.5 International Regulations for Preventing Collisions at Sea (COLREGs)

The International Regulations for Preventing Collisions at Sea were created by the International Maritime Organization (IMO) in 1972 and became effective in 1977. They are a detailed collection of regulations intended to avoid collisions among ships. The laws known as COLREGs are internationally accepted and applied in open seas and navigable waters connected to the sea. The COLREGs comprise 38 rules categorized into five divisions, accompanied by four annexes outlining further requirements. The regulations encompass numerous areas of navigation, such as vessel behavior in various conditions, the use of lights and shapes to indicate vessel status, and sound signaling for transmitting maneuvers.

7.5.1 Key Aspects of COLREGs

- **Conduct of Vessels in Any Condition of Visibility (Rules 4–10)**: Specifies how vessels should act in fog, darkness, and other situations where visibility is compromised.
- **Lights and Shapes (Rules 20–31)**: Outlines the specific lighting configurations vessels must display to indicate their type and activity, enhancing identification and safety at night and during restricted visibility.
- **Sound and Light Signals (Rules 32–37)**: Defines the signals vessels must use to communicate their actions, intentions, and presence to other vessels, particularly in conditions of poor visibility.

7.5.2 Additional International Conventions

- **International Convention on Collision, 1910**: One of the earliest agreements aimed at preventing collisions at sea.
- **International Convention on Salvage and Assistance at Sea, 1910**: Establishes the legal framework for salvage operations and assistance between vessels.
- **Brussels Conventions of 1952**: Sponsored by the International Maritime Committee, these conventions further refined the rules and practices for maritime navigation and safety.

7.5.3 Application and Compliance

The COLREGs are applicable to all vessels on the high seas and in any associated navigable waterways accessible to seagoing vessels. National authorities can enforce particular regulations within their internal seas as long as these regulations align with the principles outlined in the COLREGs. The term "vessel" includes all types of watercraft, such as non-displacement vessels, WIG craft (Wing-In-Ground effect vehicles), and seaplanes, that are used or can be used for transportation on water.

7.5.4 Navigation Lights and Marks

- **Masthead Light**: A white light placed over the fore and aft centerline of the vessel.
- **Side Lights**: Green light on the starboard side and red light on the port side, indicating the vessel's direction.
- **Stern Light**: A white light placed as nearly as practicable at the stern.

- **Special Configurations**: Vessels under 20 m in length may have specific arrangements for displaying navigation lights, including options for sailing vessels and vessels under oars.

Maritime safety standards have evolved in conjunction with developments in navigation technology, leading to a notable decrease in the likelihood of collisions at sea. Vessel operators are ultimately responsible for preventing accidents by remaining vigilant and following laws to maintain safety at sea.

Mariners who understand and adhere to the COLREGs can sail confidently, ensuring they are taking the required measures to avoid collisions and promote a secure maritime setting.

7.6 Critical Elements of Navigation Essential for Offshore Operations

This chapter establishes the foundation for comprehending the crucial aspects of navigation necessary for offshore engineering operations in the oil industry. Proficiency in these principles guarantees the secure and effective functioning, a basic necessity in the investigation, advancement, and extraction of offshore oil reserves.

Integrating the intricacies of offshore operations into the basics of navigation highlights the crucial importance of navigation in the effective and safe completion of offshore engineering activities. Precise navigation and operational planning are required for various operations such as oil offloading, water and diesel transfer, cargo movement, diving operations with support vessel assistance, and coordinating simultaneous operations between platforms and support boats. Here is a summary of how navigation concepts relate to these essential offshore activities.

7.6.1 Oil Offloading Operations

Oil offloading operations are crucial for offshore oil production, as they include transferring crude oil from offshore platforms to tankers or onshore facilities through pipelines. These activities are intricate and necessitate accurate navigation and operational planning to guarantee safety and efficiency. Key navigational components for effective oil dumping operations are:

- **Dynamic Positioning (DP) Systems**: Modern tankers and offloading vessels are often equipped with DP systems that use thrusters and propellers, controlled by advanced computer algorithms, to maintain the vessel's position automatically. These systems rely on input from various sensors, including GPS, to compensate for environmental forces such as wind, waves, and currents, ensuring the vessel remains stationary during the offloading process.

- **Environmental Monitoring**: Accurate and real-time monitoring of sea currents, wind direction, and speed is crucial for planning and conducting offloading operations. Environmental conditions significantly affect the relative positioning of the production facility and the offloading vessel, requiring constant adjustments to maintain the safety and integrity of the transfer process.
- **Proximity to Other Structures**: Navigators must be acutely aware of the proximity of their vessel to other offshore structures, including platforms, rigs, and other vessels. Safe distances must be maintained to prevent collisions and ensure the integrity of the offloading equipment, such as hoses and pipelines.
- **Transfer Hoses and Pipeline Management**: The management of transfer hoses or pipelines is a critical aspect of offloading operations. Navigational planning includes the deployment and retrieval of these connections, ensuring they are not subjected to undue stress or environmental forces that could lead to disconnection or damage.
- **Emergency Response Planning**: Effective navigation during oil offloading operations includes comprehensive emergency response planning. This involves predefined procedures for quick disconnection, vessel maneuvering to avoid hazards, and coordination with support vessels and emergency response teams.
- **Communication and Coordination**: Clear and constant communication between the offloading vessel, the production facility, and any supporting vessels is essential. Coordination ensures that all parties are aware of their roles, the current status of the operation, and any adjustments required due to changes in environmental conditions or operational parameters.
- **Regulatory Compliance**: Navigators must also ensure that offloading operations comply with international regulations, such as the International Convention for the Prevention of Pollution from Ships (MARPOL), and local laws governing offshore oil and gas activities. This includes adherence to designated shipping lanes, environmental protection measures, and reporting requirements.

Successful oil offloading operations in offshore conditions require experience, advanced technology, and meticulous planning. Offshore operations can accomplish operational efficiency and environmental safety by incorporating dynamic positioning systems, environmental monitoring, and efficient communication and coordination. These navigation techniques are crucial for reducing hazards and guaranteeing the secure and effective transportation of crude oil in the challenging conditions of the open ocean.

7.6.2 Monitoring and Drift Limits

Drift is the inadvertent displacement of a vessel from its assigned location caused by environmental factors such currents, wind, and waves. Controlling and limiting drift is essential to prevent accidents, protect the equipment, and ensure the safety of all people. Below are essential factors to consider when monitoring and controlling drift limits:

- **Drift Monitoring Systems**: Advanced monitoring systems, often integrated within dynamic positioning (DP) systems, continuously assess the vessel's position and movement. These systems utilize GPS, gyrocompasses, and reference sensors to detect any drift from the predetermined position. Real-time data allows for immediate corrective action to reposition the vessel within safe limits.
- **Setting Drift Limits**: Before commencing critical operations as offloading operations, specific drift limits are established based on the operational parameters, environmental conditions, and the capabilities of the offloading and receiving vessels. These limits are determined to ensure that any movement does not compromise the safety of the operation or the structural integrity of the transfer hoses, pipelines, and connection systems.
- **Environmental Assessment**: Continuous assessment of environmental conditions is vital for drift management. Changes in wind speed and direction, sea currents, and wave height can significantly impact vessel drift. Navigational plans must be adaptable, with strategies in place to adjust operations in response to environmental changes.
- **Tandem Mooring Operations**: In some offloading operations, especially those involving shuttle tankers, tandem mooring systems are used where the tanker maintains a certain distance behind the production facility, connected by hoses. Precise navigation and drift control are essential in this configuration to maintain the correct alignment and distance between the vessels.
- **Emergency Disconnect Procedures**: In the event that drift exceeds safe limits, emergency disconnect procedures must be initiated to prevent damage to the offloading infrastructure and ensure the safety of both vessels. These procedures should be well-practiced and understood by all crew members involved in the operation.
- **Training and Simulations**: Regular training and simulation exercises for crew members on handling drift and operating DP systems are essential. These exercises should include scenarios involving equipment failure, sudden environmental changes, and emergency disconnects to prepare the crew for real-world challenges.
- **Coordination with Support Vessels**: Support vessels, such as tugboats, may be used to assist in maintaining the position of the offloading vessel or to provide immediate response in case of excessive drift. Coordination and communication with these vessels are key components of the overall drift management strategy.

By integrating these procedures into the navigation planning and execution critical operations, offshore engineering projects can greatly diminish the risks linked to vessel drift. Continuous monitoring, precise management of drift limits, and appropriate response techniques are crucial for the safe and efficient operations in the offshore environment.

7.6.3 Transfer of Water and Diesel from Support Vessels

Transferring water, diesel, and other necessary supplies from support vessels to offshore platforms is a crucial operation that requires accurate navigation and operational planning. These procedures require precise positioning of support boats in relation to the platforms and strict control over the vessel's drift to guarantee the safety and effectiveness of the transfer process. Key factors to consider for the successful transfer of water and diesel from support vessels include navigation and operational aspects.

- **Dynamic Positioning (DP) Systems**: Support vessels equipped with DP systems use a combination of thrusters and propellers, controlled by sophisticated computer algorithms, to maintain a precise position relative to the offshore platform. These systems are crucial for compensating for environmental forces such as wind, waves, and currents, ensuring the vessel remains stationary during the transfer operation.
- **Drift Control**: Effective drift control is paramount during the transfer of water and diesel to prevent any unintended movement that could lead to disconnection of transfer hoses, potential collisions, or environmental spillage. Monitoring systems integrated within the DP system continuously assess the vessel's position and detect any deviation from the set parameters, allowing for immediate corrective actions to be taken.
- **Environmental Monitoring**: Continuous assessment of environmental conditions is vital for planning and conducting transfer operations. Navigators must consider the impact of wind speed and direction, sea currents, and wave height on the vessel's drift and adjust the DP system accordingly to maintain the required positioning.
- **Alignment and Connection Procedures**: Precise navigation ensures the correct alignment of the support vessel with the offshore platform, facilitating the safe and efficient connection of transfer hoses or pipelines. This includes careful approach maneuvers, accurate positioning alongside the platform, and the secure fastening of connections to commence the transfer.
- **Emergency Disconnect and Response Procedures**: In the event of excessive drift or other emergency situations, quick and efficient procedures for disconnecting the transfer hoses or pipelines are essential to prevent damage and spillage. Support vessels and platform crews must be well-trained in these procedures and ready to respond to any incidents.
- **Communication and Coordination**: Effective communication between the support vessel, platform personnel, and other vessels in the vicinity is crucial for the coordination of transfer operations. This includes clear communication of intentions, operational status, and any adjustments required due to changes in environmental conditions or operational parameters.
- **Regulatory Compliance and Safety Measures**: Adherence to international and local regulations governing the transfer of hazardous materials, such as diesel, is mandatory. This includes compliance with safety standards for the handling

and transfer of fuels, emergency preparedness, and environmental protection measures.

Utilizing latest navigation systems, efficient drift control techniques, and strict operational regulations guarantees the safe and successful delivery of water and diesel from support vessels to offshore platforms. These procedures are crucial for reducing hazards linked to such operations, safeguarding the environment, and guaranteeing the uninterrupted functioning of offshore installations.

7.6.4 Movement of Cargoes Between Platforms and Support Boats

The transportation of cargoes, such as equipment, supplies, and personnel,[2] between offshore platforms and support boats is a crucial aspect of offshore operations that necessitates careful navigational planning and implementation. The safety and effectiveness of this operation rely on accurate vessel positioning, ideal timing, and efficient coordination between platforms and support boats. Here are important factors to consider for navigation and the significance of drift control in guaranteeing the safe transportation of cargoes:

- **Dynamic Positioning (DP) Systems**: Support boats equipped with DP systems utilize advanced technology to maintain a precise position relative to the offshore platform. These systems automatically adjust the vessel's thrusters based on data from GPS, gyrocompasses, and environmental sensors, countering the effects of wind, waves, and currents to stabilize the vessel during cargo operations.
- **Drift Control Measures**: Controlling the drift of the vessel is crucial during cargo movement operations to prevent accidental collisions, ensure the safety of personnel involved in the transfer, and avoid damage to cargo. Drift control is achieved through continuous monitoring of the vessel's position and immediate adjustments via the DP system to counteract any unwanted movement caused by environmental forces.
- **Environmental Monitoring**: Successful cargo movement operations require real-time monitoring of environmental conditions, including wind speed and direction, sea currents, and wave height. This information allows navigators to anticipate changes that could affect the vessel's stability and positioning, enabling proactive adjustments to maintain safety.
- **Precise Docking Procedures**: Navigation aids, along with DP systems, facilitate precise docking procedures, allowing support boats to approach and maintain their position alongside the platform safely. This precision is vital for the safe loading and unloading of cargoes, particularly in adverse weather conditions where the risk of drift is increased.

[2] In the case of transporting personnel, additional safety rules must be considered in accordance with current standards and legislation and the best available safety practices.

- **Coordination and Communication**: Effective communication and coordination between the platform and the support boat are paramount. This includes pre-operation planning, real-time communication during cargo movement, and clear signals for starting and completing the transfer. Coordination ensures that both parties are synchronized, minimizing the risk of accidents and ensuring the operation's efficiency.
- **Emergency Procedures**: In case of excessive drift or other emergencies, well-established procedures for quickly securing or disconnecting cargo are necessary. Both platform and support boat crews must be trained in these emergency procedures to respond effectively, ensuring the safety of personnel and the integrity of the cargo.
- **Regulatory Compliance**: Adherence to international and national regulations governing offshore cargo movement is mandatory. This includes safety standards for cargo handling, personnel transfer, and environmental protection measures to prevent pollution.

It is crucial to include advanced navigation systems, efficient drift control, and strict coordination protocols to ensure the safe and efficient transportation of cargoes between offshore platforms and support boats. These techniques reduce hazards related to cargo transfer operations, assuring the uninterrupted and secure functioning of offshore facilities.

7.6.5 *Diving Operations with Support Vessel Assistance*

Diving activities, essential for maintaining, inspecting, and repairing underwater structures, necessitate accurate navigational tactics to safeguard divers. Vessels are outfitted with advanced navigation and communication systems to support operations, allowing precise monitoring of divers' whereabouts, efficient coordination of surface support activities, and safe deployment and recovery of divers. Key navigational factors and the significance of drift control are crucial for guaranteeing the safety of diving operations.

- **Advanced Navigation Systems**: Support vessels utilize advanced navigation systems, including dynamic positioning (DP) and underwater tracking technologies, to maintain a precise location above the dive site. These systems allow for real-time adjustments to the vessel's position, ensuring stability and accuracy in diver deployment and retrieval.
- **Drift Control**: Controlling the vessel's drift is paramount during diving operations to prevent moving away from the designated dive site, which could endanger divers by increasing the distance for their return to the surface or exposing them to hazardous areas. Drift control is achieved through continuous monitoring and adjustments by the DP system, counteracting environmental forces such as wind, waves, and currents.

- **Environmental Monitoring**: Accurate assessment of surface and underwater currents, visibility, and proximity to potential hazards is essential for planning and conducting safe diving operations. Environmental conditions directly affect divers' safety and the operation's overall success, requiring adjustments to navigational plans and dive procedures.
- **Communication Systems**: Robust communication systems are vital for maintaining constant contact between the divers, support vessel, and dive operation coordinators. These systems facilitate the immediate relay of information regarding divers' status, environmental conditions, and any necessary adjustments to the operation, ensuring rapid response to any issues that may arise.
- **Safety Protocols and Emergency Procedures**: Comprehensive safety protocols and emergency procedures must be in place to address potential risks, including diver distress, adverse environmental conditions, or equipment failure. These protocols include predefined actions for diver recall, emergency ascent, and medical emergencies, supported by the precise navigational capabilities of the support vessel to quickly locate and assist divers.
- **Coordination with Other Vessels**: Navigational planning for diving operations includes coordinating with other vessels in the vicinity to ensure they are aware of the diving activity and maintain a safe distance. This minimizes the risk of interference or accidents that could jeopardize the divers' safety.
- **Regulatory Compliance**: Diving operations must comply with international and national regulations governing underwater work, including safety standards for commercial diving, environmental protection measures, and vessel operation. Adherence to these regulations ensures the protection of divers and the marine environment.

Applying advanced navigation and communication systems, efficient drift control, and strict safety regulations is crucial for the successful implementation of diving operations with support vessel assistance. These methods guarantee the safety and effectiveness of underwater maintenance, inspection, and repair tasks, protecting divers and maintaining the integrity of offshore infrastructure.

7.6.6 Coordinating Simultaneous Offshore Operations

Efficiently managing multiple offshore operations simultaneously, like oil offloading, supply transfers, and diving activities, requires a complex navigational approach involving real-time vessel tracking, effective communication, and seamless integration of navigation systems. Coordination is essential for organizing movements and activities to prevent interference and preserve safety and efficiency in offshore operations. Here are crucial factors to consider when overseeing such intricate operations:

- **Advanced Navigation and Dynamic Positioning Systems**: Utilizing advanced navigation and dynamic positioning (DP) systems is crucial for maintaining precise control over the position and drift of each vessel involved in simultaneous operations. These systems allow for the accurate station-keeping of vessels relative to each other and the offshore platform, minimizing the risk of collision and ensuring the safe transfer of personnel, supplies, and equipment.

- **Drift Control**: Effective drift control is paramount in the coordination of simultaneous offshore operations. It involves continuous monitoring and adjustment of each vessel's position to counteract environmental forces such as wind, waves, and currents. Drift control is essential to prevent unintended movement that could compromise the safety of operations, particularly when multiple activities are occurring in close proximity.
- **Real-Time Environmental Monitoring**: Continuous assessment of environmental conditions is vital for adapting operational plans and navigational strategies in real-time. This includes monitoring wind speed and direction, sea currents, and wave height, which can significantly impact vessel positioning and the safety of operations.
- **Integrated Communication Networks**: Establishing robust and integrated communication networks among vessels, platforms, and operation coordinators ensures that all parties remain informed of current operations, environmental conditions, and any changes to planned activities. Effective communication is key to coordinating simultaneous operations, allowing for rapid adjustments and the synchronization of activities.
- **Situational Awareness and Advanced Planning**: Maintaining high levels of situational awareness and engaging in advanced planning are critical for identifying potential conflicts between operations before they occur. This involves the use of predictive modeling and scenario planning tools to anticipate and mitigate risks associated with conducting multiple operations simultaneously.
- **Emergency Preparedness and Response**: Comprehensive emergency preparedness and response plans are essential for addressing potential incidents that may arise during simultaneous operations. These plans should include protocols for emergency disconnection, evacuation, and rescue operations, tailored to the complexities of conducting multiple activities concurrently.
- **Regulatory Compliance and Best Practices**: Adhering to international and national regulations governing offshore operations, as well as industry best practices, ensures that simultaneous activities are conducted safely and responsibly. Compliance with regulations such as the International Convention for the Safety of Life at Sea (SOLAS) and the International Management Code for the Safe Operation of Ships and for Pollution Prevention (ISM Code) is essential for protecting personnel, the environment, and assets.

Conclusively, effectively managing simultaneous offshore operations necessitates a comprehensive strategy for navigation, highlighting the significance of modern navigation systems, drift control, environmental monitoring, and integrated communication. By focusing on these crucial components, offshore operations can attain both safety and efficiency, which are vital for the effective exploration, development, and extraction of offshore oil reserves.

Bibliography

1. M. Blewitt, *Celestial Navigation for Yachtsmen*, 13th edn. (Adlard Coles Nautical, London, 2016)
2. N. Bowditch, *The American Practical Navigator: An Epitome of Navigation* (National Imagery and Mapping Agency, Bethesda, 2002)
3. T.J. Cutler, *Dutton's Nautical Navigation*, 15th edn. (Naval Institute Press, Annapolis, 2003)
4. P.R.V. Lobo, *Meteorologia e Oceanografia: Usuário Navegante*, 4th edn. (Vozes, Petrópolis, 2019)
5. Marinha do Brasil, Diretoria de Hidrografia e Navegação, *International Chart Series: INT 1: Symbols, Abbreviations and Terms Used on Charts*, 5th edn. (DHN, Niterói, 2022). https://www.marinha.mil.br/chm/sites/www.marinha.mil.br.chm/files/u1974/carta12000.pdf. Accessed 20 Apr 2025. Known as Chart 12.000, published by Diretoria de Hidrografia e Navegação, Marinha do Brasil (2022)

Chapter 8
Legal Compliance

Abstract The chapter "Legal Compliance" presents the historical evolution and current framework of international and national maritime regulations governing offshore oil production. It begins with the origins of maritime classification, tracing back to Edward Lloyd's Coffee House in 17th-century London, which led to the creation of the Lloyd's Register—the world's first ship classification society. The text then outlines the establishment of Bureau Veritas and the global movement toward standardized safety practices following the Titanic disaster, culminating in the International Convention for the Safety of Life at Sea (SOLAS) in 1914 and the later creation of the International Maritime Organization (IMO) in 1948. The chapter explains the purpose and structure of key international conventions—SOLAS, MARPOL, ISM Code, and the MODU Code—highlighting their essential roles in ensuring vessel safety, environmental protection, and operational management in the maritime industry. It also details Brazil's adherence to these conventions through national legislation, notably LESTA (Law No. 9,537/1997) and the NORMAM regulations issued by the Directorate of Ports and Coasts (DPC) of the Brazilian Navy, which govern offshore operations and maritime safety in national waters. The chapter concludes by describing the Master Table of an Offshore Installation, which defines crew roles and minimum staffing requirements—including the Offshore Installation Manager (OIM) and other key positions—emphasizing that compliance with both international and national regulations is fundamental to the safety, environmental responsibility, and operational integrity of offshore oil production facilities.

In 1648, Edward Lloyd was born in London and would go on to become a significant figure in the history of maritime commerce. He opened the Lloyd's Coffee House on Tower Street in 1686, and this location would serve as a popular meeting place for sailors, merchants, and owners where business was discussed and information about navigation was exchanged.

As time went on, in December 1691, Lloyd's Coffee House was moved to Lombard Street, where it continued to thrive. Between 1696 and 1697, Edward Lloyd made

M. Vinicius da Silva Neves et al., *Offshore Operations Engineering*,
SpringerBriefs in Applied Sciences and Technology,
https://doi.org/10.1007/978-3-032-04715-1_8

the decision to start the Lloyd's News: a publication that reported on maritime and safety activities and further benefited the maritime community.

The popularity of the Lloyd's Coffee House led to the establishment of the Lloyd's List in 1734, a publication that was vital to the industry and provided details about ships and their conditions. Later, in 1760, a Register Society was established based on this publication, becoming the first organization (or society) to practice classification related to ships, focusing in the quality of their hulls and equipment. Later, this group would give rise to the Lloyd's Register Society, whose goal was to establish private standards that, when followed, would ensure safety and quality in shipbuilding.

By publishing the first edition of the Register of Ships in 1764, Lloyd's Register established a unique and independent classification system that assessed the condition of ships as well as their hull and equipment. As the nineteenth century progressed, Brazil gained independence in 1822 and gave birth to its own Marinha, marking a significant milestone for the nation.

In Antwerp, the Bureau Veritas Classification Society (BV) was established in 1828, marking a significant advancement in maritime classification. A few years later, in 1834, the Lloyd's Register group was reorganized, becoming the Classification Society in the manner we know nowadays and has released its first rules for the classification of vessels.

In 1850, Norwegian insurance societies adopted common rules for building ships, reflecting the growing need for universal protection in the maritime industry. The maritime scene changed drastically after the tragic RMS Titanic sinking on April 14, 1912, which claimed the lives of over 1500 people, and in response to this disaster, the international maritime community realized how vital it was to come together and put strong measures in place to ensure maritime safety. In this way, the government sponsored the first international conference on the preservation of human life at sea, which was held in London in 1913, and the convention received its first approval in January 1914.

By 1948, the growing complexity of maritime transportation and the need for a coordinated approach to maritime safety regulation led to the establishment of the International Maritime Organization (IMO). At UN (United Nations Organization) conference held in Geneva, nations from all over the world came together with the goal of creating an organization capable of creating and advancing international standards for maritime safety, environmental protection, and maritime transportation efficiency. The IMO became an essential forum for international cooperation, facilitating the development of conventions and regulations aimed at ensuring the safety and sustainability of international waters.

The IMO updated and adopted the SOLAS convention in 1960, reflecting the evolving needs for maritime safety.

Today, the SOLAS Convention stands as one of the three most significant international instruments, which regulates issues related to maritime safety and pollution prevention, being unquestionably the most significant convention in the field of maritime transportation. The other two conventions are the MARPOL and STCW Conventions.

Recognizing the significance of these rules, Brazil joined the IMO in 1963 and joined the organization's council in 1967, taking an active role in the governance of global maritime issues. Thus, the history of maritime commerce and naval safety continued to evolve, shaping the future of navigation.

Due to its worldwide nature, the maritime sector needs a universal standard to guarantee the protection of marine life, reduce pollution, and promote efficient and safe shipping. The International Convention for the Safety of Life at Sea (SOLAS) and the International Maritime Organization (IMO) both have a role to play in this.

The international shipping industry's performance in terms of safety, security, and the environment is governed by IMO, a specialized agency of the United Nations. Its primary duty is to provide a fair, practical, broadly embraced, and uniformly applied regulatory framework for the maritime sector.

Countries that are members of the IMO are obligated to adhere to its conventions, by their maritime authorities, and this include SOLAS and others important conventions.

8.1 SOLAS

The current SOLAS has been updated and amended on numerous occasions, and nowadays includes various chapters related to the safety of shipping, including:

1. Chapter I—General Provisions: Includes regulations concerning the survey of the various types of ships and the issuing of documents signifying that the ship meets the requirements of the Convention.
2. Chapter II-1—Construction—Subdivision and Stability, Machinery and Electrical Installations: The subdivision of passenger ships into watertight compartments must be such that after assumed damage to the ship's hull the vessel will remain afloat and stable.
3. Chapter II-2—Fire Protection, Fire Detection and Fire Extinction: Fire safety provisions for all ships with detailed measures for passenger ships, cargo ships and tankers.
4. Chapter III—Life-Saving Appliances and Arrangements: Life-saving appliances and arrangements, including requirements for life boats, life jackets and other life-saving appliances.
5. Chapter IV—Radiocommunications: The Global Maritime Distress and Safety System (GMDSS). All passenger ships and all cargo ships of 300 gross tonnage and upwards on international voyages are required to carry radio equipment, including satellite Emergency Position Indicating Radio Beacons (EPIRBs) and Search and Rescue Transponders (SARTs).
6. Chapter V—Safety of Navigation: This chapter sets out provisions requiring all ships to carry nautical charts and nautical publications to plan and display the ship's route for the intended voyage and to plot and monitor positions throughout

the voyage. An up-to-date collection of nautical publications is required to be carried on board to aid in the safe navigation of the ship.

7. Chapter VI—Carriage of Cargoes: Requirements for the stowage and securing of all types of cargo and cargo containers except liquids and gases in bulk.
8. Chapter VII—Carriage of Dangerous Goods: The carriage of dangerous goods in packaged form and in solid form in bulk, except for the carriage of oil, gas and bulk chemicals, which is covered by other IMO instruments.
9. Chapter VIII—Nuclear Ships: Regulations for nuclear-powered ships.
10. Chapter IX—Management for the Safe Operation of Ships (ISM Code): Requires the ship's owner and the company operating the ship to develop, implement and maintain a Safety Management System.
11. Chapter X—Safety Measures for High-Speed Craft: Code of safety for high-speed craft.
12. Chapter XI-1—Special Measures to Enhance Maritime Safety: Various regulations on enhancing maritime safety.
13. Chapter XI-2—Special Measures to Enhance Maritime Security: Includes the International Ship and Port Facility Security Code (ISPS Code).
14. Chapter XII—Additional Safety Measures for Bulk Carriers: Specific requirements for bulk carriers over 150 m in length.

8.2 IMO

A specialized UN organization tasked with controlling shipping is the International Maritime Organization (IMO). The primary responsibility of the IMO is to create and uphold a comprehensive regulatory framework for shipping. Today, it is also responsible for safety, environmental issues, legal issues, technical cooperation, maritime security, and shipping efficiency.

It was originally founded in Geneva in 1948 as the Inter-Governmental Maritime Consultative Organization (IMCO) under the supervision of the United Nations. The objective was to establish a permanent organization to ensure international cooperation between governments regarding regulations and practices related to the technical issues affecting shipping engaged in international trade; to encourage the adoption of high standards in maritime safety, navigational efficiency, and the prevention of marine pollution from ships. The International Maritime Organization (IMO) took over from the IMCO and began operating in 1959. By creating and upholding a comprehensive regulatory framework, the IMO continues to enhance maritime safety and security and prevent pollution from ships.

Key IMO Conventions include:

1. International Convention for the Safety of Life at Sea (SOLAS), 1974, as amended
2. International Convention for the Prevention of Pollution from Ships (MARPOL), 1973, as amended

3. Convention on the International Regulations for Preventing Collisions at Sea (COLREG), 1972
4. International Convention on Load Lines (LL), 1966
5. International Convention on Standards of Training, Certification and Watchkeeping for Seafarers (STCW), 1978, as amended
6. Convention for the Suppression of Unlawful Acts Against the Safety of Maritime Navigation (SUA), 1988, as amended
7. International Convention on Maritime Search and Rescue (SAR), 1979
8. International Convention for the Control and Management of Ships' Ballast Water and Sediments (BWM), 2004
9. International Convention on the Control of Harmful Anti-fouling Systems on Ships (AFS), 2001
10. International Convention for Safe Containers (CSC), 1972
11. Convention on Facilitation of International Maritime Traffic (FAL), 1965
12. International Convention on Salvage (SALVAGE), 1989
13. Nairobi International Convention on the Removal of Wrecks (NAIROBI WRC), 2007
14. Hong Kong International Convention for the Safe and Environmentally Sound Recycling of Ships (HKC), 2009.

8.3 International Safety Management (ISM) Code

A global guideline for the safe management and operation of ships, emphasizing security and pollution avoidance. The International Maritime Organization (IMO) approved it in 1993, and the SOLAS Convention made it mandatory in 1998.

The ISM Code specifies safety management goals and requires the establishment of a safety management system (SMS) by the ship owner or ship manager. The SMS includes instructions and procedures for critical shipboard operations related to the safety of the crew and the vessel, as well as protocols for crisis management.

According to the Code, the corporation must also establish and document each employee's role, authority, and relationship to all tasks related to safety and pollution control while ensuring the provision of appropriate resources and shore-based support.

Every corporation is required to appoint a person or persons with direct access to top management onshore. A copy of the Safety Management Manual, which contains the processes mandated by the Code, ought to be stored on board.

8.4 MARPOL

The International Convention for the Prevention of Pollution from Ships (MARPOL) is the main international convention covering prevention of pollution of the marine environment by ships from operational or accidental causes. It was adopted on 2 November 1973 at IMO and covers not only accidental and operational oil pollution but also pollution by chemicals, goods in packaged form, sewage, garbage, and air pollution.

MARPOL is structured into six technical Annexes, each of which regulates a distinct source of pollution and all are mandatory except for Annex IV.

The Annexes are:

1. Annex I—Regulations for the Prevention of Pollution by Oil
2. Annex II—Regulations for the Control of Pollution by Noxious Liquid Substances in Bulk
3. Annex III—Prevention of Pollution by Harmful Substances Carried by Sea in Packaged Form
4. Annex IV—Prevention of Pollution by Sewage from Ships
5. Annex V—Prevention of Pollution by Garbage from Ships
6. Annex VI—Prevention of Air Pollution from Ships.

Each Annex of MARPOL has its own set of regulations that require different systems and equipment to be installed on board, and different operational procedures to be followed.

8.5 MODU

The Mobile Offshore Drilling Units Code, or MODU Code, is a set of international regulations established by the International Maritime Organization (IMO) to ensure the safety and integrity of mobile offshore drilling units. These vessels are designed for drilling operations to explore or exploit resources beneath the seabed, including oil, gas, and minerals.

The MODU Code was first adopted in 1979 in response to the rapid evolution of design technology for mobile offshore drilling units and the introduction of new features to improve technical and safety standards. However, the 1979 MODU Code was superseded by the 1989 MODU Code due to several tragic MODU casualties that highlighted the need for a review of the international safety standards established by the IMO.

The 1989 MODU Code was later updated to the 2009 MODU Code, which is currently in force. The 2009 MODU Code was adopted to incorporate amendments to the Convention on International Civil Aviation that affected the 1989 Code's provisions for helicopter facilities, as well as amendments to SOLAS regulations referenced in the 1989 Code.

The 2009 MODU Code applies to mobile offshore drilling units, the keels of which are laid or at a similar stage of construction on or after January 1, 2012. The Code establishes design criteria, construction standards, and safety measures for mobile offshore drilling units, aiming to minimize risks to the units, personnel on board, and the environment.

The Code encompasses a wide range of safety measures, including structural integrity, stability, machinery, electrical installations, fire safety, life-saving appliances, and escape arrangements. It also includes provisions for the operation of helicopter facilities and for units that utilize dynamic positioning systems.

The MODU Code is continually reviewed and updated as necessary to reflect advancements in design and technology, ensuring that it remains relevant and effective in maintaining the safety and integrity of mobile offshore drilling units. Key safety and environmental rules outlined in the MODU Code include:

1. **Design and Construction**: The code provides detailed guidelines for the design and construction of mobile offshore drilling units, including the materials used, the structural integrity of the units, and the equipment installed on them.
2. **Operational Safety Measures**: The code outlines various safety measures to be followed during the operation of the units. This includes safety procedures for drilling operations, safety equipment to be carried on board, and measures for the safety of personnel on board.
3. **Environmental Protection Standards**: The code also includes provisions for minimizing the environmental impact of the operation of mobile offshore drilling units. This includes measures for the prevention of oil spills, the disposal of waste, and the reduction of emissions.
4. **Inspection and Certification**: The code requires that mobile offshore drilling units undergo regular inspections and surveys to ensure compliance with the code's provisions. These inspections are carried out by officers of the Administration or by organizations recognized by it.

The MODU Code is enforced by the IMO and the administrations of the countries where the units are registered. Compliance with the code is mandatory for all mobile offshore drilling units operating internationally.

Let's now move on to the Brazilian legislation.

8.6 LESTA

The Brazilian Law for the Security of Waterway Traffic (LESTA) was established by Law No. 9537 of December 11, 1997. It provides the legal framework for the safety of waterway traffic in waters under national jurisdiction. It covers a wide range of activities, including the operation of offshore production facilities. Here are some key points:

1. **Scope**: The law applies to all vessels, whether national or foreign, in Brazilian waters. It also applies to Brazilian vessels in international waters or the waters of other countries, unless otherwise stipulated by international agreements.
2. **Authority**: The Brazilian Navy is responsible for implementing the law and has the authority to inspect vessels, impose penalties, and take other necessary measures to ensure safety.
3. **Safety Measures**: The law outlines various safety measures, including the requirement for vessels to be properly equipped and maintained, the need for crew members to be adequately trained, and the obligation to follow navigation rules.
4. **Penalties**: The law provides for penalties in case of non-compliance, including fines, suspension or cancellation of the vessel's registration, and even imprisonment in some cases.
5. **Investigation of Accidents**: The law also provides for the investigation of maritime accidents and incidents, with the aim of preventing future occurrences.

8.7 NORMAM

The Directorate of Ports and Coasts (DPC), part of the Brazilian Navy, has issued a set of rules known as the Norms of the Brazilian Maritime Authority (NORMAM). These standards govern the operation of offshore production facilities as well as other marine activities.

Boats used for port support, offshore support, and offshore special activities are subject to NORMAM-01 restrictions. It covers a wide range of topics, including the design, setup, functionality of these boats, and the skills required for their crew.

NORMAM-04 offers guidelines for the management of offshore platforms and associated facilities. It addresses issues such as the planning, construction, hardware, systems, setup, use, and maintenance of these platforms. The guidelines also include provisions for the health and safety of workers operating these platforms.

The NORMAM-01 and NORMAM-04 guidelines provide specific instructions for the operation of offshore industrial facilities in accordance with Brazilian legislation. All individuals involved in these operations must have access to these guidelines.

In conclusion, the operation of offshore manufacturing facilities is governed by a complex web of international and national laws and regulations. These laws and rules work to safeguard the marine environment, the integrity of the facilities, and the safety of the workers engaged. Compliance with these rules and regulations is essential for ethical business practices. Key aspects of the NORMAM regulations that are relevant to offshore oil production facilities include:

1. **Safety Standards**: The NORMAM regulations set out detailed safety standards for offshore oil production facilities. These cover areas such as the design and construction of facilities, the maintenance and inspection of equipment, and the training and certification of personnel.

2. **Environmental Protection**: The regulations include measures to prevent pollution from offshore oil production activities. These include requirements for waste management, the prevention of oil spills, and the monitoring of environmental impacts.
3. **Emergency Response**: The NORMAM regulations require offshore oil production facilities to have emergency response plans in place. These plans should cover potential scenarios such as oil spills, fires, and other accidents, and should include measures to protect personnel and minimize environmental damage.
4. **Inspections and Compliance**: The regulations provide for regular inspections of offshore oil production facilities to ensure compliance with safety and environmental standards. Non-compliance can result in penalties, including fines and the suspension of operations.

8.8 Master Table of an Offshore Installation

The minimum crew requirements and roles for offshore oil production facilities can vary depending on the specific regulations of the governing body and the nature of the operation. However, there are several key roles that are typically required for the safe and efficient operation of these facilities. One of these roles is the Offshore Installation Manager (OIM).

The Offshore Installation Manager (OIM) is the most senior manager of an offshore platform operating on a shift basis. The OIM is in charge of the installation, the personnel onboard, and the day-to-day management of operations. The OIM is responsible for ensuring compliance with local and international regulations, and for ensuring the safety and efficiency of operations.

Key roles on an offshore oil production facility might include:

1. **Offshore Installation Manager (OIM)**: In charge of the installation, the personnel onboard, and the day-to-day management of operations.
2. **Maintenance Supervisor**: Responsible for overseeing the maintenance of mechanical and electrical equipment on the platform.
3. **Drilling Supervisor**: In charge of all drilling operations and associated activities.
4. **Safety Officer**: Responsible for monitoring and improving the safety performance of the platform.
5. **Production Supervisor**: Oversees the process of oil and gas production.
6. **Medic**: Provides medical care for personnel on the platform.
7. **Crane Operator**: Operates the cranes for lifting and moving equipment.
8. **Control Room Operator**: Monitors the production process from a central control room.
9. **Deck Crew**: Carry out general operations and maintenance tasks.
10. **Radio Operator**: Operates and maintains all aspects of the installation's onboard communication systems.

These roles are typically included in the Master Table of the offshore installation, which outlines the roles, responsibilities, and minimum qualifications of the personnel onboard. The exact composition of the crew can vary depending on the size, type, and operational requirements of the facility.

Please note that this is a general overview and may not cover all the specific requirements for the operation of offshore oil production facilities. For a detailed understanding of these requirements, it would be advisable to consult the original documents or seek legal advice.

Bibliography

1. Brasil, Lei nº 9.537, de 11 de dezembro de 1997. Dispõe sobre a segurança do tráfego aquaviário em águas sob jurisdição nacional e dá outras providências (1997). http://www.planalto.gov.br/ccivil_03/leis/l9537.htm. Accessed 21 Apr 2025
2. International Maritime Organization, *International Convention for the Prevention of Pollution from Ships* (1973). https://www.marinha.mil.br/ccaimo/marpol. Accessed 15 Apr 2025
3. Netherlands Regulatory Framework, MODU Code, 2009—code for the construction and equipment of mobile offshore drilling units, 2009 (2009 MODU Code) (Resolution A.1023(26)) (2009). https://puc.overheid.nl/nsi/doc/PUC_2379_14/2/. Accessed 15 Apr 2025
4. *Comissão Coordenadora para os Assuntos da Organização Marítima Internacional* (CCA-IMO, 2025). http://www.ccaimo.mar.mil.br/. Accessed 20 Apr 2025
5. *Convenção Internacional para Salvaguarda da Vida Humana no Mar* (Índice, 2025). https://www.marinha.mil.br/ccaimo/solas. Accessed 15 Apr 2025
6. International Maritime Organization, *The International Safety Management (ISM) Code* (2025). https://www.imo.org/en/ourwork/humanelement/pages/ISMCode.aspx. Accessed 15 Apr 2025
7. Marinha do Brasil, Diretoria de Portos e Costas, *Normas da Autoridade Marítima para Embarcações Empregadas na Navegação em Mar Aberto*. NORMAM-01/DPC (2005). https://www.marinha.mil.br/dpc/sites/www.marinha.mil.br.dpc/files/normam01_0.pdf. Accessed 18 Apr 2025
8. Marinha do Brasil, Diretoria de Portos e Costas, *Normas da Autoridade Marítima para Operação de Embarcações Estrangeiras em Águas Jurisdicionais Brasileiras*. NORMAM-04/DPC (2013). https://www.marinha.mil.br/dpc/sites/www.marinha.mil.br.dpc/files/bibliografia/qt/NORMAM-04-DPC.pdf. Accessed 18 Apr 2025

Chapter 9
Life Cycle Management

Abstract The chapter "Life Cycle Management" discusses the importance of strategically managing the aging and life extension of offshore oil production facilities to ensure operational safety, economic efficiency, and asset longevity. It explains that proper aging management prevents premature failures—mainly those caused by corrosion and fatigue—while extending the productive lifespan of installations beyond their initial design, thus maximizing asset value and deferring costly decommissioning. The chapter introduces the concept of the "bathtub curve", relating failure probability to cost, downtime, and risk, and proposes monitoring facility performance and economic outcomes as key indicators of the onset of aging. Once this decline is detected, operators can revise integrity strategies, plan refurbishments, or replace equipment to sustain performance and profitability. It emphasizes that each facility should conduct annual aging evaluations after reaching midlife to anticipate degradation and optimize investment decisions for life extension. Complementary tools—such as Structural Reliability Analysis (SRA), Quantitative Risk Assessment (QRA), and Integrity Management—are presented as essential for maintaining safety and efficiency under the ALARP principle, supported by technologies like IoT, predictive analytics, and digital twins. The chapter also covers Life Cycle Structural Integrity Management, highlighting design criteria, maintenance, and sustainability practices that ensure resilience against environmental and accidental events, and introduces Design for Structural Damage Tolerance, focused on preventing progressive failure through resilient design, advanced simulations, and training. Overall, it concludes that effective aging assessment and integrated integrity management are critical for ensuring the safe, profitable, and sustainable operation of offshore oil production facilities throughout their life cycle.

M. Vinicius da Silva Neves et al., *Offshore Operations Engineering*,
SpringerBriefs in Applied Sciences and Technology,
https://doi.org/10.1007/978-3-032-04715-1_9

9.1 General Overview of Aging and Life Extension

Managing the life cycle is crucial for the strategic operation of offshore oil production installations. Efficiently managing the aging process of structures, systems, and equipment can greatly prolong the operating lifespan of a facility beyond its original estimates. This prolongation is economically beneficial since oil fields can still produce economically viable reserves. Additionally, it avoids the necessity of making new investments in production systems and the expenses related to decommissioning and reconnecting wells. Moreover, it enables the operator to incorporate extended reserves into their asset portfolio, so increasing their value alongside established reserves.

Poor aging management can cause premature failure modes, especially those related to corrosion, making life extension impractical and possibly requiring early decommissioning. Ensuring the integrity of structures, systems, and equipment is essential for operational safety, as a well-maintained facility guarantees the effectiveness of preventative and risk mitigation measures.

As facilities get older, their structures, systems, equipment, and especially their risk barriers deteriorate. An asset integrity strategy needs to identify and resolve these challenges to uphold an acceptable risk level. Figure 9.1 illustrates the "bathtub curve," which shows how offshore production facilities behave over time by linking failure probability with parameters including costs, shutdowns, and operational risks.

The aging of installations is determined by both their operational environment and the level of usage. A mechanism is required to predict the beginning of aging, considering the uncertainty in these elements. Recognizing the beginning of the aging process enables the development of new maintenance and integrity tactics to address failure modes associated with age, fatigue, corrosion, and degradation.

The HSE's KP4 program in Great Britain proposes assessing the overall performance of installations as an indicator of approaching the aging phase. Economic

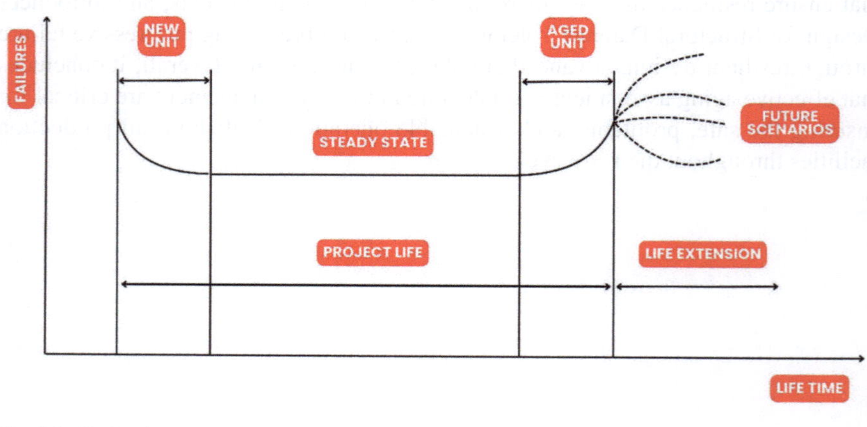

Fig. 9.1 Bath tub curve

results, which are strongly linked to production efficiency, also act as a valuable measure. Throughout the steady state portion of the bathtub curve, production and economic results should stay at elevated levels. A consistent decline in these parameters signifies the beginning of aging.

Anticipating the onset of the aging process and forecasting economic outcome trends allows operators to calculate the viable investment needed to restore the initial performance. Consistent reassessment and forecasting of these elements are crucial for thorough analysis and accurate prediction.

Each installation should perform this examination annually once it reaches its midpoint to precisely evaluate the beginning of aging. This proactive method enables rapid adjustments in integrity strategy, refurbishment, or equipment replacement to uphold performance, economic outcomes, and profitability.

Aging evaluation is crucial for evaluating the possibility of extending life and selecting the best time to start life extension initiatives. This method reduces production and economic losses caused by the emergence of new failure modes resulting from age and degradation.

9.2 Aging and Life Extension Analysis

As already mentioned, a significant issue in aging is the possible deterioration of components and systems that are crucial for maintaining preventive and risk-reducing safety measures. Installations with compromised physical integrity often pose significant concerns to operational safety. Operators must strengthen risk monitoring systems and guarantee the effectiveness of these crucial safety measures.

The graph depicts the "bathtub curve" (Fig. 9.2), which shows the performance of offshore production facilities over their lifespan. The y-axis can depict various factors such as costs, shutdowns, loss of production, inefficiency, operational risks, increased manpower needed to operate and maintain an aging facility, and heightened demand for critical resources for system maintenance. These factors worsen as the probability of system failure increases.

As discussed above, a simple analysis can determine the aging of production facilities based just on their ages. However, the complexity arises from the interaction between the medium in which the installation is placed and the form and intensity with which it operates, which can alter the aging curve based on the pressures experienced by the facility and its equipment over time. Some elements have a random nature that hinders a direct assessment to identify the point on the "bathtub curve" or anticipate when the probability of failure starts to increase. We believe that aging is determined by the equipment's state and how it changes over time, rather than just its age. Therefore, it is essential to create a framework for forecasting to account for uncertainties and the onset of aging in facilities, a phase where the likelihood of failure rises.

The goal is to recognize when the installation or its crucial structures and systems begin to deteriorate, leading to a higher rate of failure. Identifying the onset of the

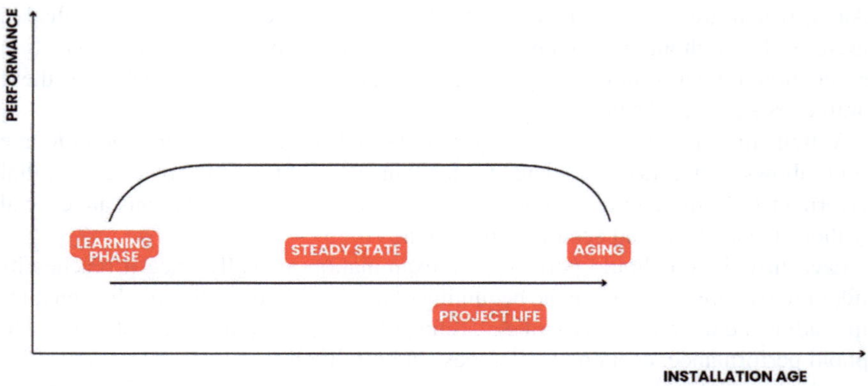

Fig. 9.2 Opposite bath tub curve

aging phase enables the establishment of new maintenance and integrity measures to prevent various failure modes associated with aging, fatigue, corrosion, and degradation commonly observed during this period. Once we identify that the installation is entering or will enter the aging phase, it becomes easier to introduce new integrity strategies such as equipment substitution, refurbishment, or adequation. This is possible because there is more time available and, consequently, more oil revenues to financially support the strategy adjustments. Looking to the installation as single equipment, it is easier to follow its general performance (production efficiency) instead of the general failure rate. That´s why we propose to look to the bath tub curve from the opposite perspective, substituting the failure rate by the installation's general performance, as shown in Fig. 9.2.

The same proposal was done by HSE (Healthy and Safety Executive) in Great Britain on its program HSE KP4—Key Program 4 (Aging and life extension 2010–2014), where the overall installation's performance is suggested to be the aspect to be used to evaluate if the installation is entering the aging phase of the bath tub curve, as it is presented in Fig. 9.3.

Other aspect that can be used to identify when the installation enters the aging phase is the installation's economic outcomes that has direct correlation with its production performance. In other words, during the steady state phase of the bath tub curve, the production performance shall maintain its maximum level, as the economic outcomes as well. And when it starts to continuously decrease, the economic outcomes shall follow the same path, as we can see in Fig. 9.4.

If you are able to predict when the aging phase has begun or will begin and to project the curve of economic outcomes, you know how much you can invest to restore the original performance. In order words, all the economic revenues that are lost due to the reduction in production performance (increasing in failure rates) over time and brought to present value represent the money amount available to restore the installation's original performance.

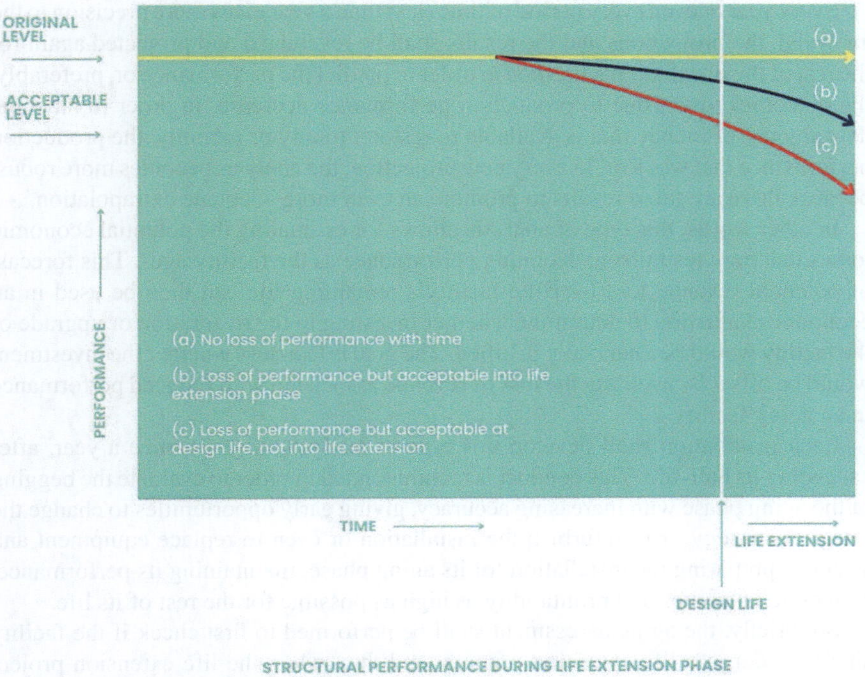

Fig. 9.3 Approach of aging evaluation by HSE KP4—Key Program 4 (aging and life extension 2010–2014)

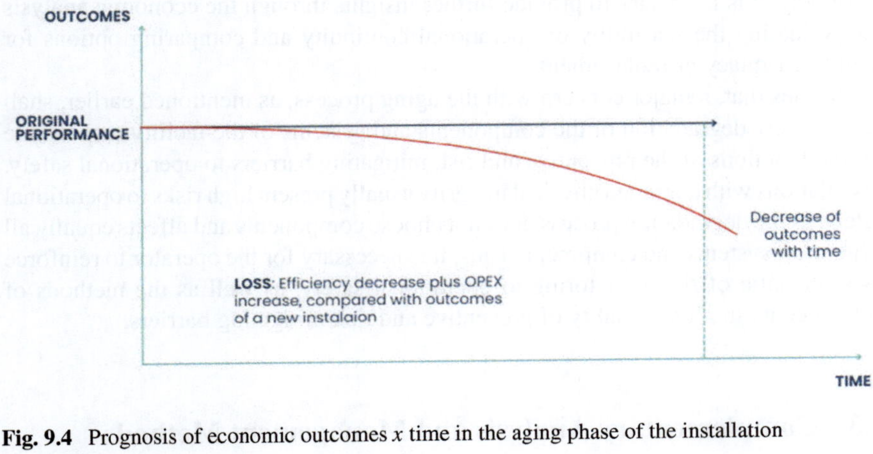

Fig. 9.4 Prognosis of economic outcomes *x* time in the aging phase of the installation

Every year or even every period of time (less than a year gives more precision to the analysis), the projections and the results shall be revaluated and projected again for the rest of the installation's lifetime in order to predict the performance or, preferably, the outcomes losses due to production performance decrease, in order to have the new amount of money that is available to restore, totally or partially, the production performance that was lost. In every new projection, the analysis becomes more robust because there are more results to promote an even more adequate extrapolation.

In other words, this type of analysis allows for estimating the potential economic losses that may result from declining performance as the facility ages. This forecast of potential revenue loss over the facility's remaining life can then be used in an economic evaluation to determine whether investing in the restoration or upgrade of the facility would be financially justified. The goal is to assess whether the investment would be offset by avoiding the loss of revenue associated with reduced performance in an aging facility.

Each installation shall develop this type of analysis, at least once a year, after exceeding its half-life. This conduct is recommended in order to evaluate the begging of the aging phase with increasing accuracy, giving early opportunities to change the integrity strategy or to refurbish the installation or even to replace equipment and systems, preparing the installation for its aging phase, maintaining its performance, economic outcomes and profitability as high as possible for the rest of its life.

So, briefly, the aging assessment shall be performed to first check if the facility should undergo a life extension process, and then when the life extension project should begin to make the Production and consequently economic losses, that are consequences of new failure modes that arise with the degradation and aging of the installation's structures, systems and equipment, as minimum as possible.

Finally, it is necessary to provide further insights through the economic analysis for evaluating the feasibility of operational continuity and comparing options for facility adequacy or replacement.

Besides that, a major concern with the aging process, as mentioned earlier, shall be the likely degradation of the components and systems of the facility responsible for the functions of the preventive and risk mitigating barriers to operational safety. Installations with degraded physical integrity usually present high risks to operational safety, as the degradation process does not choose components and affects equally all structures, systems and equipment. Thus, it is necessary for the operator to reinforce its systematic of risk monitoring to operational safety, as well as the methods of guaranteeing the functionality of preventive and risk mitigating barriers.

9.3 Complementary Analysis and Management Methods

Other management tools and methods of analysis are improving rapidly and can improve the integrity and aging management to reduce the efforts related to life cycle management and life extension. Below we can see a summary of some of these methods and tools:

1. **Structural Reliability Analysis and Quantitative Risk Assessment**: According to a study on offshore platforms in the Gulf of Thailand, a quantitative risk assessment (QRA) and structural reliability analysis (SRA) are crucial for evaluating and predicting platform risk development. This approach helps in optimizing inspection strategies, thereby prolonging inspection intervals, especially for well-designed and well-maintained platforms. This method aligns with the principle of keeping operation risks as low as reasonably possible (ALARP) while minimizing inspection costs and resources.
2. **Integrity Management as a Pillar of Efficiency**: Another study emphasizes the role of integrity management as a fundamental pillar in maximizing efficiency for offshore production units. It advocates for a life cycle approach to integrity management, encompassing material and technology selection, preservation, inspection, and maintenance. Preventive mechanisms for evaluating and detecting structural and functional degradations are essential, considering the operational and environmental conditions to which offshore facilities are exposed.
3. **Life Cycle Structural Integrity Management**: An overview of structural integrity management highlights the importance of controlling risks related to extreme environmental events, structural degradation, and accidental events. Adequate design criteria, inspection, repair, maintenance, and quality assurance are key measures. A quantitative design approach, particularly for dealing with crack degradation phenomena, is suggested for a life cycle approach to structural integrity management.
4. **Design for Structural Damage Tolerance**: Focusing on the design for structural damage tolerance, research suggests developing operational design standards for accidental collapse limit states. This involves ensuring an acceptable safety level against progressive failure due to operational errors, fabrication errors, and abnormal deterioration during operation. The concept of damage tolerance is crucial for achieving safety through inspection, monitoring, and repair strategies.

We will go into further detail on each of the upcoming sections.

9.3.1 *Structural Reliability Analysis and Quantitative Risk Assessment*

Quantitative risk assessment and structural reliability analysis are essential to offshore platform lifecycle management. These techniques follow the ALARP principle and extend inspection intervals for well-maintained and designed platforms by optimizing inspection methodologies. This strategy minimizes inspection-related expenses and resource use while reducing operating hazards in an efficient manner.

To further emphasize the significance of quantitative risk assessment (QRA) and structural reliability analysis (SRA) in the lifecycle management of offshore platforms, it's critical to understand how these approaches work with cutting-edge technical tools and data analytics. A more comprehensive understanding of potential failure mechanisms and their effects is made possible by SRA and QRA, which integrate complex modeling approaches with probabilistic assessments. This is especially important in settings with extremely changeable and difficult operating conditions.

Moreover, these analyses are not static, they are dynamic processes that gain from ongoing data collection and real-time oversight. On offshore platforms, the integration of sensor technologies and IoT (Internet of Things) has created new opportunities for gathering enormous amounts of data about operational parameters, environmental conditions, and structural health. This data can be used to make proactive decisions by proactively addressing possible risks before they become significant problems by using the SRA and QRA lenses to assess it.

The significance of human variables in risk assessment is another important consideration. Despite being highly technical, SRA and QRA still need to take operational decision-making, human error, and maintenance procedures into account. The competence and training of employees are key factors in guaranteeing the dependability of offshore structures. Operators can create more comprehensive strategies that cover both the technical and human aspects of offshore operations by integrating human factors into these assessments.

Furthermore, fulfilling international safety standards and adhering to regulations depend on SRA and QRA. They give operators a formalized way to show their dedication to environmental and safety management, which is crucial in a world where sustainability and corporate responsibility are closely monitored.

In summary, Structural Reliability Analysis and Quantitative Risk Assessment are more than just tools for risk management; they are integral components of a holistic approach to offshore platform lifecycle management. By embracing technological advancements, continuous learning, human factors, and regulatory compliance, these methodologies not only enhance safety and efficiency but also contribute to the long-term sustainability and viability of offshore operations.

9.3.2 Integrity Management

Integrity management is clearly essential to efficiency in offshore production units, as was covered in the preceding section. It includes the selection of materials and technology, preservation, inspection, and maintenance, with a focus on a lifespan approach. This comprehensive approach is essential given the variety of operational and environmental difficulties offshore facilities confront. The lifespan and effectiveness of these installations are consequently guaranteed by preventive processes, which are essential in identifying and assessing structural and functional deteriorations.

Examining Integrity Management in greater detail as the basis for productivity in offshore production units requires considering this approach's complexity. Integrity management encompasses more than just the choice of materials and technology, preservation, inspection, and upkeep; it also entails a strategy alignment with resilience to climate-related concerns and environmental sustainability. It becomes essential to incorporate climate adaptability into design and operating strategies for offshore facilities since they frequently operate in harsh and unpredictable conditions. This includes being aware of the effects of extreme weather events and being ready for them.

Integrating digital technology, such as predictive analytics and digital twins, is another essential component of integrity management. Operators can simulate multiple situations and forecast the results of different tactics with the help of digital twins, which are virtual representations of physical assets. This allows for more informed decision-making. On the other side, predictive analytics makes use of both past and current data to identify possible problems before they arise, enabling proactive maintenance and risk reduction.

Integrity management is more effective when industry members work together and share expertise. Operators might adopt creative strategies and steer clear of typical problems by studying the experiences and best practices of others. This cooperative approach includes collaborating closely with regulatory agencies, making sure that the most recent safety and environmental regulations are followed, and helping to create stronger industry standards.

Moreover, effective Integrity Management depends on personnel development and training. In addition to improving operational safety, making sure staff members are knowledgeable about the newest techniques and technology promotes an innovative and continuous improvement culture. Realizing the full potential of procedural and technological breakthroughs in Integrity Management frequently depends on this human component.

In conclusion, integrity management is a broad and developing subject in offshore production units. Advanced technology, environmental awareness, industrial cooperation, regulatory compliance, and ongoing workforce development must all be balanced. Offshore operators can guarantee the durability, efficiency, and sustainability of their facilities in an increasingly complex and demanding operational setting by addressing these disparate but interconnected factors.

9.3.3 Life Cycle Structural Integrity Management

Conversely, structural deterioration, extreme environmental events, and unintentional events all carry risks that must be mitigated through the use of Life Cycle Structural Integrity Management. The significance of suitable design requirements, frequent inspections, prompt maintenance and repairs, and strict quality assurance is emphasized by this management strategy. It is advised to use a quantitative design approach to guarantee structural integrity throughout the facility's existence, especially when addressing phenomena like crack degradation.

To further elaborate on the notion of Life Cycle Structural Integrity Management, it is imperative to underscore the significance of innovative materials and inventive engineering approaches in augmenting the robustness of offshore structures. The key to prolonging the life of these structures is the application of corrosion-resistant metals and composite materials to, for instance, adjust to changing environmental conditions. These materials not only provide increased durability, but they also lessen the need for frequent and intensive maintenance.

Apart from advancements in materials, the utilization of state-of-the-art technology like as computational fluid dynamics, finite element analysis, and 3D modeling is vital to the initial design and continuous evaluation of structural integrity. With the aid of these technologies, engineers can now simulate environmental effects, load capabilities, and possible failure locations with greater accuracy and foresight, leading to more accurate and prescient structural design and modification.

Working together with academic institutions and business associations is essential for the advancement of Life Cycle Structural Integrity Management. The sector can remain up to date on the most recent advancements and optimal methodologies by means of collaborative research endeavors and mutual learning environments. This cooperative strategy encourages creativity and aids in establishing new standards for the maintenance of structural integrity.

Moreover, the life cycle management approach must incorporate sustainability and environmental care. This include preparing for ultimate decommissioning in a way that minimizes ecological damage and considering the environmental impact of the materials and processes used in construction and maintenance.

In conclusion, life cycle structural integrity management is a dynamic and diverse field that calls for a combination of modern materials, innovative technology, meticulous maintenance plans, cooperative research, and environmental awareness. In the face of constantly changing operational and environmental problems, the offshore industry may efficiently manage risks and guarantee the longevity and safety of its structures by embracing these features.

9.3.4 Design for Structural Damage Tolerance

A crucial field of study is "Design for Structural Damage Tolerance", which focuses on creating operational design guidelines for limit situations of unintentional collapse. This entails preserving a reasonable degree of safety against incremental failures brought on by mistakes in operation, defects in fabrication, and unusual deterioration while in use. Achieving safety requires a strong understanding of damage tolerance, which calls for effective systems for inspection, monitoring, and repair.

Investigating the incorporation of resilience-based design concepts is essential. This method goes beyond conventional safety precautions by considering a structure's capacity to tolerate unanticipated circumstances and carry on operations even if a system component fails. Resilience-based design ponders the structure's full lifecycle, from inception to decommissioning, making sure it can adjust and bounce back from harm to preserve operational integrity throughout time.

In this field, the application of sophisticated simulation tools and machine learning techniques is also essential. With the use of these technologies, possible damage scenarios and their effects on structural integrity may be more precisely modeled. Engineers can better understand how structures will perform in extreme situations and design accordingly to reduce these risks by simulating different damage and failure scenarios.

To advance the area of Design for Structural Damage Tolerance, industry, academics, and regulatory agencies must work together. These kinds of partnerships may result in the creation of new norms and regulations that take into account the most recent findings and developments in technology. They also make it easier to share lessons learned and best practices, which is crucial in a field where worker safety is of the utmost importance.

Furthermore, it is imperative to consider the human element in both design and operation. It is important to use simulation-based training tools and training programs to make sure staff members are equipped to handle and respond to structural failures and damages. This entails being aware of the constraints placed on the structures they operate with as well as the appropriate procedures for upkeep, inspection, and emergency response.

To sum up, structural damage design in order to fully address the complex topic of tolerance, it is necessary to combine resilient design principles, advanced technology, creative materials, cooperative research, and a significant emphasis on human issues. The offshore industry may greatly improve the safety and dependability of its structures by including these components, making sure they are prepared to withstand the rigors of hostile maritime conditions.

To sum up, aging assessment plays a critical role in deciding if life extension initiatives in offshore oil production facilities are feasible and when they should be implemented. This evaluation reduces output and financial losses brought on by new failure modes that develop as a result of aging and degradation. It is crucial to combine structural reliability analysis with quantitative risk assessment and integrity management over the entire lifecycle. Maintaining good performance, financial results, and profitability of offshore oil production facilities depends on these coordinated efforts.

Bibliography

1. Á. Angulo, G. Edwards, S. Soua et al., Mooring integrity management: novel approaches towards in situ monitoring, in *Structural Health Monitoring—Measurement Methods and Practical Applications*, ed. by M. Rivas-Lopez, W.F. Fuentes, O. Sergiyenko (InTech, Croatia, 2017), pp. 87–108. https://doi.org/10.5772/intechopen.68386
2. Health and Safety Executive, *Key Programme Final Reports* (2008). https://www.hse.gov.uk/offshore/programmereports.htm. Accessed 15 Apr 2025

3. Health and Safety Executive, *Key Programme 4 (KP4): Ageing and Life Extension Programme* (2013). https://www.hse.gov.uk/offshore/assets/docs/kp4-report.pdf. Accessed 15 Apr 2025
4. Health and Safety Executive, *Maintaining the Integrity of Offshore Installations* (2021). https://www.spe-aberdeen.org/wp-content/uploads/2021/12/DAY-2-HSE-TOPSIDESUK2021MBfinal2.pdf. Accessed 18 Apr 2025
5. T. Moan, Integrity management of offshore structures with emphasis on design for structural damage tolerance. J. Offshore Mech. Arct. Eng. **142**(3), 031104 (2020). https://doi.org/10.1115/1.4045373
6. National Offshore Petroleum Safety and Environmental Management Authority, *Ageing Assets and Life Extension* (2021). https://www.nopsema.gov.au/sites/default/files/documents/2021-07/A783718.pdf. Accessed 20 Apr 2025
7. Offshore Engineer, *Protecting Assets Through Better Data Trending* (2014). https://www.oedigital.com/news/455016-protecting-assets-through-better-data-trending. Accessed 19 Apr 2025
8. Protected Pool, *Home* (2025). https://pp.one/. Accessed 18 Apr 2025

Chapter 10
Pre-operation, Operation and Maintenance

Abstract The chapter Pre-operation, Operation and Maintenance covers the essential phases to ensure the safe and efficient operation of offshore production facilities: pre-operation, which includes team formation and training, development of operational standards, maintenance process structuring, risk management, obtaining certifications, and equipment commissioning, preparing the facility to start production; and operation and maintenance, which involves monitoring safe operations, treating anomalies, managing maintenance and operational integrity, responding to emergencies, managing changes, conducting operations, and continuously evaluating management, integrating safety practices, preventive maintenance, change control, and performance analysis to ensure reliability, efficiency, and sustainability throughout the facility's lifecycle.

10.1 Pre-operation

An offshore oil production installation must complete a number of preliminary tasks before it can begin to operate in order to ensure a safe and effective operation. These activities include team formation and training, operational standard development, maintenance process structuring, understanding and managing operational safety risks, obtaining required certifications and documentation, and finally accepting, commissioning, and pre-operation of equipment. The essential components of the pre-operation stage of offshore oil production installations will be described in this chapter.

Team Formation and Training

The first step in preparing for operation is the formation and training of teams. This involves identifying the necessary roles and responsibilities, recruiting suitable personnel, and providing them with the necessary training to perform their duties effectively and safely. Training should be comprehensive, covering all aspects of operation, maintenance, safety procedures, and emergency response. It is important to emphasize to the teams that, due to the inhospitable and complex environment,

M. Vinicius da Silva Neves et al., *Offshore Operations Engineering*,
SpringerBriefs in Applied Sciences and Technology,
https://doi.org/10.1007/978-3-032-04715-1_10

working offshore requires knowing how to work as a team and providing support to others whenever necessary.

Development of Operational Standards

Operational standards provide a framework for how activities should be conducted. They establish guidelines for routine operations, maintenance procedures, safety protocols, and emergency response plans. These standards should be developed in accordance with industry best practices and regulatory requirements, and considering aspects related to the feasibility of maneuvers in relation to how they are described and what is encountered in practice in the field.

Structuring of the Maintenance Process

A well-structured maintenance process is essential for the safe and efficient operation of offshore oil production installations. This involves developing a maintenance schedule, identifying necessary maintenance tasks, and establishing procedures for routine inspections and repairs. It is extremely important to show the teams that there needs to be a routine for monitoring maintenance, where it is possible to identify points for improvement in the process, and also involve specialists when something unusual is identified.

Understanding and Managing Operational Safety Risks

Understanding and managing operational safety risks is a critical aspect of pre-operation preparation. This involves conducting risk assessments to identify potential hazards, evaluating the likelihood and severity of these hazards, and implementing measures to mitigate these risks. The Guidelines for Risk Based Process Safety from the Center for Chemical Process Safety (CCPS) provide a comprehensive framework for managing process safety risks. These documents should be known to the crew, generating awareness about the impacts of certain actions on process safety.

Certifications and Documentation

Before operation can commence, all necessary certifications must be obtained, and all required documentation must be in place. This includes certifications of compliance with safety and environmental regulations, as well as documentation of operational procedures, maintenance schedules, safety protocols, and emergency response plans.

Acceptance, Commissioning, and Pre-operation of Equipment

The final step in the pre-operation phase is the acceptance, commissioning, and pre-operation of the equipment. This involves conducting thorough inspections and tests to ensure that all equipment is functioning correctly and safely, and that it meets all regulatory requirements. Once all equipment has been accepted and commissioned, pre-operation activities can commence.

The flow below provides a step-by-step guide to the pre-operation process, ensuring that all necessary preparations are made for a safe and efficient operation.

1. **Team Formation and Training**
   - Identify necessary roles and responsibilities
   - Recruit suitable personnel
   - Conduct comprehensive training (operations, maintenance, safety procedures, emergency response)
2. **Development of Operational Standards**
   - Identify necessary operational procedures
   - Develop standards in accordance with industry best practices and regulatory requirements
   - Document and communicate these standards to all personnel
3. **Structuring of the Maintenance Process**
   - Develop a maintenance schedule
   - Identify necessary maintenance tasks (routine inspections, repairs, etc.)
   - Establish procedures for conducting these tasks
4. **Understanding and Managing Operational Safety Risks**
   - Conduct risk assessments to identify potential hazards
   - Evaluate the likelihood and severity of these hazards
   - Implement measures to mitigate these risks
   - Follow the Guidelines for Risk Based Process Safety from the CCPS
5. **Certifications and Documentation**
   - Obtain necessary certifications (compliance with safety and environmental regulations)
   - Prepare required documentation (operational procedures, maintenance schedules, safety protocols, emergency response plans)
6. **Acceptance, Commissioning, and Pre-operation of Equipment**
   - Conduct thorough inspections and tests of all equipment
   - Ensure all equipment meets regulatory requirements
   - Accept and commission all equipment
   - Conduct pre-operation activities to verify system functionality
7. **Start of Operation**
   - Once all the above steps are completed, and the installation has been verified to be ready, operation can commence.

In conclusion, the pre-operation phase is a critical period that lays the foundation for a safe and efficient operation of offshore oil production installations. By carefully planning and executing these preparatory activities, operators can ensure that their installations are well-prepared to begin operation.

10.2 Operation and Maintenance

In order to ensure safe and effective production, offshore oil production installations must be operated and maintained. Based on the Guidelines for Risk Based Process Safety from the Centre for Chemical Process Safety (CCPS), this chapter will go into detail about the process flow for these tasks: Monitoring Safe Operation, Treating Anomalies, Managing the Maintenance and Integrity Process, Responding to Emergencies, Managing of Change, Conducting Operations, and Evaluating and Reviewing Management are the stages that make up the process flow.

Monitoring Safe Operation

This stage involves a set of practices to be observed by the workforce, as guided and required by the Risk Manager. It includes communicating relevant facts such as operational situations, accident scopes, audits, and any information deemed relevant by the installation. It also involves managing changes to ensure that modifications to a process do not introduce unknown risks. The change management system will outline the steps for identifying the change, assessing the introduced risks, hierarchical authorization level, and documentation update.

Treating Anomalies

This stage involves identifying, acting upon, containing, analyzing, and correcting deviations, incidents, and accidents that occur. It is crucial to have a systematic approach to handle anomalies to prevent them from escalating into more significant issues that could potentially disrupt operations or pose safety risks.

Managing Maintenance, Integrity, and Operational Safety

Effective management of maintenance and integrity processes is fundamental to ensuring operational safety. This involves the careful oversight of safety barriers, maintaining them within their pre-established levels to guarantee reliability and effectiveness. Regular inspections, maintenance activities, and timely upgrades are essential to preserving the functionality and safety of these barriers, safeguarding the overall integrity of the installation. By integrating these practices, the maintenance and integrity process not only upholds operational safety but also ensures the resilience and long-term performance of the system.

Responding to Emergencies

This stage involves maintaining readiness and responding to emergencies. It is crucial to have a well-developed emergency response plan and the necessary resources to implement it effectively when required.

Managing of Change (MOC)

Change management is a critical aspect of maintaining safe and efficient operations in offshore oil production installations. Changes can include modifications to equipment, procedures, personnel, materials, or the operating environment. Each change carries potential risks, and it's crucial to manage these changes effectively to prevent the introduction of unknown risks.

The change management process typically involves the following steps:

1. **Identification of Change**: Any proposed change, whether it's a change in equipment, procedures, or personnel, should be identified and documented. This includes who proposed the change, what the proposed change is, and why the change is necessary.
2. **Risk Assessment**: Once a change has been proposed, a risk assessment should be conducted to identify any potential risks associated with the change. This should include both the likelihood and the potential impact of each identified risk.
3. **Review and Approval**: The proposed change and its associated risk assessment should be reviewed by a designated authority. This authority will decide whether to approve the change, reject it, or request further information or modifications.
4. **Implementation**: If the change is approved, it should be implemented in a controlled manner. This includes communicating the change to all affected personnel and providing any necessary training.
5. **Verification and Follow-Up**: After the change has been implemented, it should be verified to ensure that it has been carried out as planned. Any problems or unexpected outcomes should be addressed promptly. A follow-up review should be conducted to assess the effectiveness of the change and identify any lessons learned.

This change management process helps ensure that changes are managed in a systematic and controlled manner, reducing the risk of unexpected problems or accidents. It's an essential part of maintaining safe and efficient operations in offshore oil production installations.

Conducting Operations

This stage involves verifying the workforce's adherence to procedures, plans, and regulations. It includes compliance verification with the procedure, safe behavior verification through audits application, adherence verification to technical regulations, quality of management verification, and the level of the installation and workforce's response to emergency situations in the response plan.

The process includes:

1. **Compliance Verification**: This involves ensuring that all operations are conducted in accordance with established procedures. It's essential to consider human factors such as understanding, competence, and workload, which can impact compliance.
2. **Safe Behavior Verification**: Using tools like the audit application, safe behaviors can be monitored and reinforced. This includes considering factors such as fatigue, stress, and motivation, which can influence behavior.
3. **Adherence Verification to Technical Regulations**: This involves ensuring that all operations comply with technical regulations. Human factors such as training, competence, and communication can impact the ability to adhere to these regulations.

4. **Quality of Management Verification**: This involves assessing the effectiveness of management practices, including leadership, communication, and the management of change. These aspects can significantly impact the human factors in operations.
5. **Emergency Response Verification**: This involves assessing the level of the installation and workforce's response to emergency situations in the response plan. Human factors such as training, stress management, and teamwork can significantly impact the effectiveness of emergency responses.

By considering human factors in each of these areas, the operation of offshore oil production installations can be made safer and more efficient.

Evaluating and Reviewing Management

This stage involves maintaining readiness to identify situations and review risk management, including reviewing risk understanding. It is crucial to continually evaluate and review management practices to identify areas for improvement and implement necessary changes.

In conclusion, the operation and maintenance of offshore oil production installations involve a systematic process that ensures safe and efficient production. By following the stages outlined in this chapter, operators can effectively manage their installations and mitigate potential risks.

The flow below provides a step-by-step guide to the operation and maintenance process, ensuring that all necessary preparations are made for a safe and efficient operation.

1. **Monitoring Safe Operation**
   - Observe workforce practices as guided by the Risk Manager
   - Communicate relevant operational situations, accident scopes, audits
   - Manage changes to ensure no introduction of unknown risks
2. **Treating Anomalies**
   - Identify deviations, incidents, and accidents
   - Act upon and contain these anomalies
   - Analyze and correct the anomalies
3. **Managing the Maintenance and Integrity Process**
   - Manage the installation's maintenance and integrity
   - Maintain operational safety barriers within pre-established levels
4. **Responding to Emergencies**
   - Maintain readiness for emergencies
   - Respond effectively to emergencies as per the emergency response plan
   - Carry out training for teams in accident scenarios

5. **Managing of Change**
    - Identify and document proposed changes
    - Conduct risk assessment for the proposed changes
    - Review and approve changes by a designated authority
    - Implement approved changes in a controlled manner
    - Verify and follow-up on the implemented changes
6. **Conducting Operations**
    - Verify workforce's adherence to procedures, plans, and regulations
    - Conduct compliance verification, safe behavior verification, adherence verification, quality of management verification, and emergency response verification
7. **Evaluating and Reviewing Management**
    - Maintain readiness to identify situations and review risk management
    - Review risk understanding and management practices
    - Implement necessary changes based on the evaluation.

Bibliography

1. Center for Chemical Process Safety, *Guidelines for Risk Based Process Safety* (Wiley, Hoboken, 2007)
2. International Organization for Standardization, *ISO 55000:2014. Asset Management—Overview, Principles and Terminology* (2014). https://www.iso.org/standard/55088.html. Accessed 16 Apr 2025

Chapter 11
Safety and Emergency Management

Abstract The chapter "Safety and Emergency Management" addresses safety and emergency management in offshore oil operations, emphasizing the identification, analysis, and mitigation of risks to personnel, installations, and the environment. It covers concepts of process and occupational safety, highlighting the role of human factors, asset integrity, environmental protection, legal compliance, and best practices. The chapter outlines hazard identification methods, such as HAZOP and FMEA, and risk mitigation strategies including fire and gas detection systems, emergency shutdowns, and preventive measures following IOGP principles. Emergency management is addressed through planning, preparation, response, and recovery for scenarios such as fires, explosions, gas leaks, marine collisions, and harsh environmental conditions. It stresses the importance of trained Emergency Response Teams (ERT), clear communication, leadership under the Incident Command System (ICS), and regular drills to ensure effective coordination, protection of life and assets, and operational continuity.

There are many sources of risk for people and installations in offshore operations. Some risks are linked to the nature of the work performed, such as falls, electric shocks, being struck by objects, slips, and finger pinching. Other risks are associated with the main purpose of the installation and the vessel itself, such as oil spills, fires, explosions, structural damages, or severe motions due to harsh weather conditions. All these risks must be identified and assessed in advance, in order to quantify them and define preventive and mitigative actions to eliminate or minimize their potential impacts.

Despite the extensive studies, procedures, and methodologies implemented to prevent accidents, the inherent risks of offshore production activities can, sometimes, lead to incidents. Therefore, additional measures must be taken to ensure an appropriate response to emergencies.

This chapter aims to present key concepts in safety engineering, discuss the main risks in an installation that can lead to significant losses, and outline preventive measures. Additionally, it describes essential aspects of emergency management, including response plans, training, drills, and methodologies.

M. Vinicius da Silva Neves et al., *Offshore Operations Engineering*,
SpringerBriefs in Applied Sciences and Technology,
https://doi.org/10.1007/978-3-032-04715-1_11

11.1 Process Safety in Offshore Operations

11.1.1 Basic Concepts

Process safety and occupational safety are two critical aspects of offshore oil production installations. They are closely linked with asset integrity, environmental protection, legal compliance, and the application of best practices. As these installations age, the importance of these aspects becomes even more pronounced due to increased risks and operational challenges.

Process Safety is a discipline that focuses on the prevention of fires, explosions, and accidental releases of hazardous chemicals. It is particularly crucial in the context of aging offshore oil production installations, where the risk of process safety incidents increases due to factors such as equipment degradation, increased maintenance needs, and changes in operating conditions. Effective process safety management can help identify these risks and implement measures to mitigate them, thereby extending the life of the platform and maintaining its operational efficiency.

Occupational Safety is concerned with the prevention of work-related injuries and illnesses. It involves creating a safe and healthy work environment by identifying, assessing, and controlling risks that may potentially cause workplace accidents and diseases.

Loss of Primary Containment (LOPC) as defined by API in its Guide to Reporting Process Safety Events, it represents "an unplanned or uncontrolled release of any material from primary containment, including non-toxic and non-flammable materials (e.g., steam, hot water, nitrogen, compressed CO_2, or compressed air)" [2].

Human Factors play a significant role in both process and occupational safety. These factors include the design of tasks, equipment, and systems, as well as the physical and mental capabilities of workers. Understanding these factors can help in designing safer work systems and procedures, thereby reducing the risk of accidents.

Asset Integrity is the ability of an asset to perform its required function effectively and efficiently while safeguarding life and the environment. The processes and activities associated with maintaining asset integrity are key to managing risks and ensuring the continued safe operation of oil production platforms.

Environmental Protection is an inherent part of safety and emergency management. It involves minimizing the environmental impact of operations, managing waste effectively, and preventing accidental releases of hazardous substances.

Legal Compliance is a fundamental aspect of safety and emergency management. It involves adhering to all relevant laws, regulations, and standards related to offshore oil production. This includes compliance with process safety regulations, occupational health and safety laws, environmental regulations, and asset integrity standards.

Best Practices in safety and emergency management involve the application of proven methods and techniques to improve safety performance. This includes the use of risk assessment tools, safety management systems, emergency response plans, and continuous improvement processes.

In essence, process safety, occupational safety, human factors, asset integrity, environmental protection, legal compliance, and the application of best practices are all interconnected aspects of safety and emergency management. Effective management of these aspects is crucial to the safe and efficient operation of offshore oil production installations, particularly as they age. This holistic approach to safety and emergency management ensures the protection of people, assets, and the environment, while also ensuring legal compliance and operational efficiency.

11.1.2 Identification and Analysis of Hazards

The nature of the processes involved in an offshore installation significantly contributes to a potentially hazardous environment for major accidents, as it typically combines the presence of flammable and polluting fluids in large volumes and at high pressures. Additionally, other risks and hazards arise from the equipment, activities, and systems necessary for these processes, such as the operation of high-power compressors and generators, storage of toxic or asphyxiating chemicals, handling of large and heavy loads, and pumping of high-temperature steam. Identifying and understanding the main hazards and risks is the first step toward effective safety management.

The occurrence of LOPC is likely the most frequently materialized risk at an offshore installation. It can lead to oil spills, which may be confined to the interior of the installation or, in a worst-case scenario, may exceed the boundaries of the installation and reach the ocean. Gas leaks are usually more dangerous because they mix with air and cannot be confined. Depending on the amount of gas, an explosion can occur if it encounters a heat source, even something as small as a spark.

One of the major disasters in recent years occurred on April 20, 2010, when an explosion took place on the Deepwater Horizon oil rig after a mixture of gas and drilling mud reached its topside. The rig was in the final phases of drilling the Macondo Well, located approximately 65 km offshore from the Louisiana coast. The incident resulted in the deaths of 11 workers, and the rig sank two days later. Failures in closing the Blow Out Preventer (BOP) positioned on the wellhead allowed the well to produce oil directly into the sea at a flow rate of up to 40,000 barrels per day. After more than three months of operations and attempts to stop the spill, the well was finally sealed, but more than 4.5 million barrels leaked into the Gulf of Mexico [5].

An explosion occurred on February 15, 2015, on the FPSO Cidade de São Mateus, following a gas leak in the pump room. The incident resulted in the deaths of 9 workers and left 26 others severely injured [6]. The FPSO was situated 40 km offshore from the coast of Espírito Santo in southeast Brazil.

These major accidents show that it is crucial to recognize that fire, explosions, and pollution can result from gas leaks and oil spills. These incidents of LOPC primarily occur due to hazards such as poor asset integrity (pipes and vessels), inadequate equipment maintenance plans (pumps, compressors, turbines, power generators), exceeding operational limits (pressure, temperature, flow rate) and poor management of change. Additionally, other hazards present on offshore installations include storage of hazardous substances, handling of heavy loads and environmental forces. Identifying these hazards and the risk to personnel, property, process performance and to the environment is the first phase of a Hazard Analysis.

In the second phase, each risk is evaluated based on its probability and consequences to classify it according to a tolerance criterion. For those risks whose probability and consequences are deemed unacceptable, actions to eliminate or mitigate the hazard or reduce the consequences of an accident are identified. These actions will be discussed in the next topic, but often involve adequate monitoring of variables, rapid identification of leaks or fires, providing relief and destinations for fluids in case of exceeding operational limits, emergency shutdown, and prompt emergency response.

A Hazard Analysis is typically a time-consuming process that involves a multidisciplinary team. The process can be facilitated when industry standards, recommended practices, government regulations, and company-specific procedures are incorporated into the design and construction of facilities. This inherent safety results from years of experience and lessons learned.

There are several methods to perform a Hazard Analysis. API RP 14J assembles "into one document useful procedures and guidelines for planning, designing and arranging offshore production facilities, and performing a hazards analysis on open-type offshore production facilities" [1, p. 38]. It details some methods typically used for offshore production installations such as "what if" analysis, HAZOP (hazards and operability), Failure Modes and Effects Analysis (FMEA) and Fault Tree Analysis (FTA).

11.1.3 Risk Mitigation, Fire, and Gas Systems

As previously mentioned, hazard analysis often involves implementing risk mitigation measures to lower the likelihood of a risk occurring or to minimize its impact.

Considering that accidents can occur when process conditions or even equipment parameters (rotation, vibration, torque, for example) exceed their limit values, a primary set of actions to reduce the probability of risk occurrence involves monitoring these conditions through detection methods (detectors and meters) associated with levels of an emergency shutdown (ESD). The sequence of automatic actions after a specific kind of detection is referred to as a cause-and-effect diagram.

The fire and gas system plays a vital role in risk mitigation by ensuring the rapid detection of fire and gas leaks. Fire alarms can be activated manually via manual call points or automatically through sensor detection. Not all detections need to trigger an alarm throughout the entire installation. The alert should be appropriate to the specific circumstance, whether it is merely a signal in the control room or a sound in a particular area of the installation. In an offshore installation, there are gas, flame, smoke, and heat detectors installed individually or in groups to provide the desired protection.

Smoke detectors are typically installed in accommodation spaces (rooms, stairways and corridors), rooms containing electrical apparatus (control room, electrical panels rooms, radio and telecommunication rooms), muster stations and ventilation air intakes for accommodation.

Storage areas of flammable or combustible products, as well as the main deck above cargo tanks, are generally covered by flame detectors. These types of detectors are often combined with temperature detectors in machinery spaces, such as fire water pump compartment, emergency generator compartment, electric power generator hood and gas compressor hood.

Although fusible plugs are not connected to the electric signal of fire detection systems, they can be considered detectors due to their ability to trigger the fire suppression system when exposed to fire. Therefore, in some areas, they can be used standalone to cover specific equipment or be combined with flame detectors in areas where hydrocarbons are likely to occur, such as riser connection areas, process areas, wellhead areas, pump rooms, and offloading areas.

Gas leak risk mitigation can be achieved by installing gas detectors. Typically, CH_4, CO_2, and H_2S are the gases that are subject to detection. The amount and location of gas detectors are determined by a gas dispersion study.

The response for detection depends on the gas concentration that reaches the detector. Hydrocarbon gas detectors (CH_4) use a percentage of the Lower Explosive Limit (LEL), which is the lowest required concentration that a flammable gas ignites if exposed to an ignition source. Usually, detection occurs, and an alarm is shown in the control room when 20% of the LEL is reached. Confirmation of a gas leak occurs when the concentration reaches 50–60% of the LEL. Activation of a specific Emergency Shutdown (ESD) level can be initiated upon confirmation by one detector. However, in areas covered by multiple detectors, activation is commonly triggered by a voting process, depending on the number of detectors in that area. For instance, in a Riser Balcony Area covered by four detectors, an ESD would occur only if the gas concentration in two detectors reaches the confirmation threshold.

As recently mentioned, in a cause-effect diagram, a fire and gas detection can lead to an ESD to prevent escalating the critical emergency at hand, whether it is a fire or a gas leak.

The response of a shutdown is proportional to the detected disturbance. The primary objective is to protect personnel, the environment, and property, while minimizing the impact on production. Therefore, Emergency Shutdown Systems (ESDs) are classified into levels, with each company having its own classification criteria. An example of ESD level classification could be:

- ESD-1: Emergency shutdown of single machine, causing its trip. An example is the shutdown of a compressor due to excessive vibration.
- ESD-2: Emergency shutdown of the process plant. This typically occurs when a process condition reaches a threshold value. For instance, an emergency shutdown may be triggered by an increase in the flare vessel's level or a high pressure in a knock-out drum.
- ESD-3: Emergency shutdown of the process plant and main power generation. In this case, an example would be a fire detected inside the hood of a main power generator. Typically, at this ESD level, the general alarm is activated, and emergency response procedures are initiated.
- ESD-4: Emergency shutdown of the entire process, main power generation, and well shut-in closing the Subsurface Safety Valves (SSSV). In this extreme case, it could be manually triggered due to the platform losing its position because of a mooring line rupture, for instance.

11.1.4 Risk Prevention

Despite all the actions to rapidly detect any gas leak or fire and responses such as shutdowns, general alarms or fire extinguish, a more effective approach is to implement measures to prevent fires from occurring in the first place. These measures could include, for instance, restricting the use of combustible materials, conducting regular housekeeping audits, exercising caution with hot works, and developing adequate maintenance plans for electrical installations and equipment.

In terms of process safety, some principles must be followed to avoid incidents. The International Oil and Gas Producers (IOGP) set out these ten principles:

- **Maintain Safe Isolation**: Certain activities require energy isolation—such as pressure and electricity—to prevent accidents. Therefore, it is mandatory to evaluate whether a task requires an isolation plan. If deemed necessary, the isolation plan must be specifically tailored to the task and based on up-to-date process safety information.
- **Walk the Line**: This principle emphasizes the importance of monitoring and verifying the condition of equipment and piping to ensure they align with their intended state, particularly before initiating a procedure or starting up operations. For example, critical items to be checked include valve positions and proper installation of plugs and drain valves.
- **Apply Procedures**: The implementation of updated procedures to perform critical tasks is essential to ensure that step-by-step activities are completed in the correct sequence and under safe conditions. Therefore, developing, discussing, and adhering to procedures should be a fundamental value for operational workers.
- **Sustain Barriers**: Critical safety equipment, critical procedures, and primary containment are key barriers that contribute to risk control within an installation. When these barriers are degraded, whether through inhibition, bypasses, malfunctions, or unavailability, the likelihood of an incident increases. To address this,

personnel must be properly trained, and robust systems should be in place to detect any weaknesses in these barriers. Prompt actions must then be taken to restore their integrity.

- **Control Ignition Sources**: Fires and explosions are among the leading causes of fatalities in the oil and gas industry. Therefore, ignition sources must be carefully managed to prevent contact with hydrocarbon releases. Precautions such as proper isolation of hot work, regular inspections of electrical equipment, and thorough consideration of zone classifications are key preventive measures to reduce the risk of process accidents.
- **Recognize Changes**: Effective management of changes within an installation is essential to ensure that all hazards and risks are thoroughly assessed, and all actions required by the change are properly implemented. Changes, whether permanent or temporary, may involve operating limits, process variables set points, materials, personnel, or equipment. Examples of necessary actions include updating documents and procedures, informing personnel about the change, modifying or implementing inspection plans, and adjusting process conditions accordingly. Workers must be trained to identify changes, assess the management process, and halt operations when faced with unexpected situations.
- **Respect Hazards**: Workers must consistently practice identifying hazards present in the offshore installation to enhance their awareness of potential risks. Beyond merely recognizing these hazards, they should also comprehend the potential consequences of incidents caused by them and take all necessary precautions to prevent such occurrences.
- **Stay Within Operational Limits**: This principle involves maintaining process conditions at a pre-established set point while adhering strictly to the defined limits.
- **Stop If the Unexpected Occurs**: When carrying out a task, following a procedure, or performing routine operations, unexpected situations may arise. It is crucial to encourage workers to pause in such instances, assess the situation thoroughly, and determine the appropriate next steps.
- **Watch for Weak Signals**: This principle highlights the importance of recognizing signals that indicate abnormal operational conditions, such as excessive vibrations, minor pneumatic air leaks, or unusual noises. Regular operator rounds, intended to stay vigilant and identify these abnormalities, play a crucial role in preventing potential process incidents.

11.2 Emergency Management

11.2.1 Planning and Preparation

Effective emergency management starts well before any triggering event. The initial phase of planning for emergency response involves identifying potential scenarios and detailing the necessary resources and actions to address them. In this context, a variety of emergencies can occur on an offshore installation, such as medical emergencies, marine collisions, helicopter crashes and harsh environmental conditions.

Assessing the effectiveness of emergency management involves evaluating the availability of necessary resources to address various scenarios. Some emergencies can be managed using resources within the facility itself. However, in cases where the emergency's magnitude surpasses internal capabilities, external resources may need to be activated. These resources include air medical rescue helicopters, firefighter boats, oil spill containment vessels, and rescue boats.

Here is an analysis of various emergency scenarios with regard to potential actions, resource requirements and possible escalation:

- **Medical Emergencies**: Injuries involving crew members are treated directly in the unit hospitals by the onboard healthcare professional. A team of first responders should be designated previously to provide first aid at the scene of the incident and transport the victim to the hospital if they are unable to walk. Additionally, given the unique nature of work on an offshore installation, a specialized team should be designated for confined space rescues or for working at heights. Planning should also account for the possibility of incidents involving multiple victims; in such cases, the need to set up a secondary treatment site should be assessed if the hospital lacks sufficient space. In certain instances, based on the severity of the injury, it may be necessary to request a medical helicopter to transfer individuals to appropriate medical facilities onshore.
- **Chemical and Oil Spills**: The loss of integrity in pipes and failures in seals, O-rings, and gaskets of equipment, vessels, and pipe joints are the most common causes of oil or chemical spills. A specifically designated team is responsible for containment and cleanup, utilizing appropriate Personal Protective Equipment (PPE) and products such as oil booms, absorbent pads and rolls, skimmers, and dispersants. In the event of major oil spills or when there is no cummings to prevent it, such as in the riser deck area, oil can reach the ocean. In these situations, oil recovery boats must be deployed to mitigate the environmental impact, and onshore teams must be prepared to act if there is a risk of oil reaching the shore. In the event of a combustible fluid spill, the incident can escalate to a fire.
- **Gas Leaks**: In such events, the primary objective should be to suspend the gas supply. Typically, upon the detection of a leak by sensors, an Emergency Shutdown (ESD) is activated, which triggers the general alarm. Once the source is suppressed, the leak is expected to cease due to the depressurization of the affected section or equipment. Given the propensity of gas leaks to escalate into fires and explosions, a firefighting team is usually mobilized to remain equipped and ready throughout the emergency. Furthermore, a dedicated team should be assigned to the control room to oversee the emergency response. Special attention must also be given to the potential risk of hazardous or flammable gas leaks entering the accommodations. In such cases, the air intake dampers should be closed immediately if this has not occurred automatically upon the detection by sensors installed near the air intake.

- **Fires and Explosions**: Typically, these incidents result from electrical failures, vessel overpressures, and spills of flammable fluids or gas leaks that come into contact with an ignition source. As a result, special precautions must be taken when performing hot work or protecting hot surfaces. In an offshore installation, the firefighting system ranges from portable extinguishers to automated systems like water deluge or CO_2 flooding. In extreme scenarios, firefighting boats may be deployed, and the event could escalate to a structural failure of the installation.
- **Marine Collisions and Structural Failures**: Although less frequent, maritime incidents have a greater potential for damage, as they can compromise the vessel's stability. Collisions usually happen during operations with support vessels, tankers, and accommodation vessels. Structural failures can result from the corrosion of primary structures, emphasizing the importance of an adequate asset integrity management program to ensure wear monitoring and repairs. Additionally, improper loading of cargo tanks, which exceeds the ship's resistance limits (shear stress and bending moment), can also lead to structural failures. Response actions aim to maintain vessel stability and protect human life. Evacuation or abandonment may be necessary if there's a risk of losing habitability or sinking.
- **Harsh Environmental Conditions**: Adverse weather conditions, such as storms and strong waves, can affect vessel stability, leading to cargo loss, structural damage, and falling objects. Preparation involves securing cargo, conducting height inspections to mitigate DROPS risks, and, in severe cases, suspending facility activities, shutting down production, and evacuating personnel.
- **Helicopter Crashes**: These incidents typically occur during helicopter landings or departures, potentially leading to personnel injuries, fires, or explosions. Consequently, a firefighting system is provided on helidecks, and a firefighting team must attend all helicopter operations.

Following the assessment of potential accidental scenarios in an offshore facility, as described above, emergency response plans are developed by multidisciplinary team. These plans detail the scenarios, the required response actions, the necessary or available resources, and the emergency response teams with their respective functions. An example of mandatory response plan required by IMO for certain vessels is the Shipboard Oil Pollution Emergency Plan (SOPEP), which describes actions to be adopted in the event of an oil pollution incident.

To maintain crew proficiency during emergencies, regular drills should be conducted for each scenario outlined in the emergency plans. These drills should incorporate a variety of situations, including different emergency sources, detection methods, consequences, resulting damages, firefighting techniques, and evacuation and abandonment procedures. This approach ensures comprehensive preparedness for a wide range of potential emergencies. According to SOLAS regulations [3], fire and evacuation ship drills must be conducted at least once a month.

The International Convention on Standards of Training, Certification and Watchkeeping for Seafarers (STCW) [4], as emended, establishes a minimum standard of competence for all personnel on board. The regulation details the standards for those enrolled in response teams, such as those responsible for survival crafts or firefighting. Even for crew members not part of special teams, the regulation sets requirements for Basic Safety Training.

11.2.2 Emergency Response

As previously mentioned, each emergency scenario demands the enrollment of specific professionals. These individuals form the Emergency Response Team (ERT), with each group assuming responsibility for a designated function. For instance, the teams are assigned to the following roles: communications, emergency source control, firefighting, lifesaving, and medical assistance.

During an emergency, each ERT group executes pre-designated tasks and assumes their assigned positions, awaiting further instructions. For example, the communications group, typically composed of radio operators, verifies the readiness of all systems, alerts firefighting vessels, cargo vessels, and helicopters near the installation about the emergency, and requests their Estimated Time of Arrival (ETA).

Some groups are composed of up to ten or fifteen professionals, demanding the assignment of a group leader. Typically, the marine superintendent, maintenance superintendent, and production/driller superintendent lead the source control teams. The health professional leads the medical assistance group, and the bosun leads the lifesaving team.

Depending on the number of People on Board (POB), some personnel may not be enrolled in any Emergency Response Team (ERT) group. During an emergency, these individuals must proceed to the muster station, where they will obtain lifesaving appliances, such as a life vest, and remain silent, awaiting instructions from the muster station coordinator. If there are multiple muster stations at the installation, each person is preassigned to a specific station and informed of it immediately upon arrival.

A paramount factor for the success of an emergency response is clear communication and coordinated action among the groups. Consequently, a commander is designated to oversee the entire emergency, establish priorities, evaluate the overall situation, and direct the actions of the ERT groups. The emergency commander is also responsible for major decisions, such as evacuation or abandonment of the installation, and for requesting external assistance from other vessels or onshore teams. The Offshore Installation Manager (OIM) is typically the Incident Commander.

Given the critical importance and far-reaching implications of decisions made by Offshore Installation Managers (OIMs) during emergencies, companies must exercise meticulous care in their selection, training, and ongoing professional development. OIMs are required to possess robust, multidisciplinary technical expertise across various domains, including maritime engineering, maintenance engineering, production engineering, meteorology, oceanography, as well as environmental and maritime regulations. Furthermore, proficiency in essential soft skills is indispensable, particularly in areas such as decision-making, prioritization, emergency management and response, effective teamwork, communication, problem-solving, risk management, and, on top of that, leadership.

Under normal working conditions, an inadequate leadership profile gradually and consistently undermines business results and team performance. While these issues can often be identified over time, emergencies create a fast-paced scenario where leadership ineptitude and poor decisions cannot be quickly detected or effectively corrected. To address this, it is crucial to establish a strong, collaborative, and secure work environment beforehand, enabling timely adjustments and real-time feedback from ERT (Emergency Response Team) leaders during the emergency.

The Incident Command System (ICS) is a methodology that has been developed and refined over the years to effectively command, control, and coordinate emergencies, crises, and disasters. It establishes clear standards to ensure proper organization of both on-scene and offshore teams, facilitates well-documented and effective communication, and promotes optimal resource management. The ICS management hierarchy is a globally recognized framework, proven to handle emergency responses ranging from minor incidents to large-scale crises. Its procedures and workflows are widely adopted by industries, private companies, and government agencies. Additionally, the ICS is particularly effective in situations where multiple departments or agencies must collaborate, enabling seamless sharing of information, resources, and logistical support.

References

1. American Petroleum Institute, *Recommended Practice for Design and Hazard Analysis for Offshore Production Facilities* (American Petroleum Institute, New York, 2019)
2. American Petroleum Institute, *Recommended Practice 754* (2021). https://www.api.org/oil-and-natural-gas/health-and-safety/refinery-and-plant-safety/process-safety/process-safety-standards/rp-754. Accessed 18 Apr 2025
3. International Maritime Organization, *International Convention for the Safety of Life at Sea (SOLAS)* (1974). https://www.imo.org/en/About/Conventions/Pages/International-Convention-for-the-Safety-of-Life-at-Sea-(SOLAS),-1974.aspx. Accessed 19 Apr 2025
4. International Maritime Organization, *International Convention on Standards of Training, Certification and Watchkeeping for Seafarers* (1978). https://www.imo.org/en/OurWork/HumanElement/Pages/STCW-Convention.aspx. Accessed 20 Apr 2025
5. R.K. Perrons, Assessing the damage caused by deepwater horizon: not just another Exxon Valdez. Mar. Pollut. Bull. **71**(1–2), 20–22 (2013). https://doi.org/10.1016/j.marpolbul.2013.03.016
6. J.E. Vinnem, FPSO Cidade de São Mateus gas explosion—lessons learned. Saf. Sci. **101**, 295–304 (2018). https://doi.org/10.1016/j.ssci.2017.09.021

Bibliography

1. American Institute of Chemical Engineers, *Process Safety in Upstream Oil and Gas* (Wiley, Hoboken, 2021). https://www.aiche.org/ccps/resources/publications/books/process-safety-upstream-oil-and-gas. Accessed 18 Apr 2025

2. American Petroleum Institute, *Offshore Safety* (2025). https://www.api.org/oil-and-natural-gas/health-and-safety/exploration-and-production-safety/offshore-safety. Accessed 15 Apr 2025
3. American Petroleum Institute, *Process Safety* (2025). https://www.api.org/oil-and-natural-gas/health-and-safety/refinery-and-plant-safety/process-safety. Accessed 20 Apr 2025
4. Center for Chemical Process Safety, *Guidelines for Hazard Evaluation Procedures* (Wiley, Hoboken, 2008)
5. F. Gimenes, R. Silva Neto, B.S. Silvestre, Sustainable operations and process safety management systems: implications for the offshore oil industry and Petrobras, in *The 5th International Workshop Advances in Cleaner Productions*, São Paulo, 20–22 May 2025 (2015). https://www.advancesincleanerproduction.net/fifth/files/sessoes/6B/5/gimenes_f_et_al_academic.pdf. Accessed 15 Apr 2025
6. Health and Safety Executive, *Process integrity* (2020). https://www.hse.gov.uk/offshore/processintegrity.htm. Accessed 15 Apr 2025
7. International Labour Organization, *Occupational Safety and Health and Skills in the Oil and Gas Industry Operating in Polar and Subarctic Climate Zones of the Northern Hemisphere* (ILO, Geneva, 2015). https://www.ilo.org/sites/default/files/wcmsp5/groups/public/@ed_dialogue/@sector/documents/publication/wcms_438074.pdf. Accessed 16 Apr 2025
8. Magellan X, *Safety in Oil and Gas Industry: Common Hazards & Prevention* (2025). https://magellanx.co/safety-in-oil-and-gas/. Accessed 16 Apr 2025
9. Process Safety Integrity, *Offshore Oil & Gas* (2025). https://processsafetyintegrity.com/incidents/industry/offshore/. Accessed 15 Apr 2025
10. Society of Petroleum Engineers, *Process Safety* (2025). https://www.spe.org/en/focus/process-safety/. Accessed 18 Apr 2025

MIX
Papier aus verantwortungsvollen Quellen
Paper from responsible sources
FSC® C105338

If you have any concerns about our products,
you can contact us on
ProductSafety@springernature.com

In case Publisher is established outside the EU,
the EU authorized representative is:
Springer Nature Customer Service Center GmbH
Europaplatz 3, 69115 Heidelberg, Germany

Printed by Libri Plureos GmbH
in Hamburg, Germany